Deforestation in the Postwar Philippines

University of Chicago Geography Research Paper no. 234

Series Editors

Michael P. Conzen
Chauncy D. Harris
Neil Harris
Marvin W. Mikesell
Gerald D. Suttles

Titles published in the Geography Research Paper series up to no. 232 are now distributed by the University of Chicago Press.

Deforestation
in the Postwar Philippines

David M. Kummer

The University of Chicago Press
Chicago and London

David M. Kummer is visiting assistant professor, Graduate
School of Geography, and research associate, George
Perkins Marsh Institute, at Clark University. He spent four
years living in the Philippines and traveling in Southeast
Asia to research deforestation and changing land use.

The University of Chicago Press, Chicago 60637
The University of Chicago Press, Ltd., London
© 1991 by The University of Chicago
All rights reserved. Published 1992
Printed in the United States of America
01 00 99 98 97 96 95 94 93 92   5 4 3 2 1
ISBN 0-226-46169-6 (paper)

⊗ The paper used in this publication meets the minimum
requirements of the American National Standard for
Information Sciences—Permanence of Paper for Printed
Library materials, ANSI Z39.48-1984.

Library of Congress Cataloging-in-Publication Data

Kummer, David M.
    Deforestation in the postwar Philippines / David M. Kummer.
        p.   cm. — (University of Chicago geography research paper :
    no. 234)
    Includes bibliographical references and index.
    1. Deforestation—Philippines. I. Title. II. Series.
SD418.3.P6K86   1992
333.75'137'09599—dc20   91-42845
                                                                CIP

For Mom and Dad

Contents

Figures

Tables

Preface

Central to the model of deforestation that will be presented is the notion of elite control of government in general and of forest resources in particular in the Philippines. Such controls have led to widespread corruption in the forest sector. However, this is a statement about the system as a whole and not about any particular individual. Nothing in this research should be construed to be a reflection of corruption or malfeasance on the part of any individual.

In addition, the existence of widespread corruption in the forest sector should not obscure the fact that numerous individuals have consistently and heroically maintained their integrity in the face of incredible odds. What successes have occurred in Philippine forestry in the postwar period are the result of the activities of these individuals. This research is dedicated to those Filipinos who have struggled over the past forty-five years to make Philippine forests serve the needs of the Filipino people rather than the privileged few.

Acknowledgments

I would first of all like to thank the members of my dissertation committee, Chi Ho Sham, Richard Primack, and Robert Batchelder, for their support over the past several years. Special thanks are due to Dr. Sham in this regard. I would also like to thank the two additional members of the examination committee, Cutler Cleveland and Foster Brown.

In the Philippines, I would particularly like to thank Dr. Adolfo Revilla (former Dean, College of Forestry, University of the Philippines at Los Baños), Ricardo M. Umali (Undersecretary, Department of Environment and Natural Resources), and members of the Philippine-German Forestry Resources Inventory Project (Conrado V. Gulmatico, Jurgen Schade, and Konrad M. Uebelhor). Without their assistance, the present research would not have been possible.

Of the numerous people who have assisted me in one way or another, I would especially like to mention: Heidi Arborleta (National Economic Development Authority), Romeo Bruce (University of the Philippines, Diliman), Joe Cabanayan (National Mapping and Resources Information Authority), Roger Concepcion (Bureau of Soils), M. Concepcion and Wilfrido Cruz (University of the Philippines, Los Baños), Cristina David (International Rice Research Institute), Dan Doeppers (University of Wisconsin), Leonie Fernandez (National Irrigation Authority), Lilo Gallagher (Boston University), Delfin Ganapin (University of the Philippines, Los Baños), Rene Garrucho (South Cotobato Foundation), Tim Herrin (Boston University), Robert Huke (Dartmouth College), Jeanne F. Illo (Institute of Philippine Culture), Remigio Mercado (National Economic Development Authority), Harold Olofson (University of San Carlos), Cayetano Paderanga (University of the Philippines, Diliman), Theo Panayotou (Harvard University), Percy Sajise (University of the Philippines, Los Baños), Marian Segura-de los Angeles (Philippine Institute for Development Studies), Elis Tanlapco (Dole Philippines), Curtis

Woodcock (Boston University), and Imelda Zosa-Feranil (University of the Philippines, Diliman).

A special thanks is due Alan Grainger of Stirling University (England) for his extensive written comments on a draft of my dissertation. However, I accept full responsibility for any remaining errors.

Funding for fieldwork in the Philippines (1987–88) was provided by a Fulbright dissertation fellowship and I want to thank everyone at the Philippine American Educational Foundation (Manila) for their generous support while I was there, particularly Juan Francisco and Alex Calata.

Abbreviations

A&D   Alienable and disposable
AAC   Annual allowable cut
ADB   Asian Development Bank
APST   Agricultural Policy and Strategy Team
BAS   Bureau of Agricultural Statistics
BF   Bureau of Forestry
BFD   Bureau of Forest Development
DENR   Department of Environment and Natural Resources
FAO   Food and Agricultural Organization
FDC   Forestry Development Center
FMB   Forest Management Bureau
FRCM   Forest resource condition map
ILO   International Labour Office
IUCN   International Union for the Conservation of Nature
LISREL   Linear structural relations
NCSO   National Census and Statistics Office
NEC   National Economic Council
NEDA   National Economic Development Authority
NIA   National Irrigation Authority
NRMC   Natural Resource Management Center
PCA   Principle components analysis
PFS   *Philippine Forestry Statistics*
P-GFI   Philippine-German Forestry Resources Inventory Project
PICOP   Paper Industries Corporation of the Philippines
PREPF   Population, Resources, Environment and the Philippine Futures
SSC   Swedish Space Corporation
UNEP   United Nations Environment Programme
UNICEF   United Nations Children's Fund
USAID   United States Agency for International Development

1

Introduction

Deforestation has been rapid and widespread in the Philippines, now one of the most severely deforested countries in the tropics. The rationale for a comprehensive quantitative examination of deforestation in this one country is straightforward: such research provides valuable insights into the process of forest removal in the tropical regions; and, with the exception of the present study, no such research exists. In the case of the Philippines, in contradistinction to many tropical countries, enough data exist at the provincial level that a meaningful statistical analysis for the country as a whole is possible. My ultimate goal is to construct a model of deforestation that will lead to a better understanding of this complex process.

Major Issues Raised by Tropical Deforestation

The literature on tropical deforestation is extensive and growing rapidly. In addition, tropical deforestation in the past ten years or so has received a great deal of attention from the press and major international organizations such as the World Bank. The recent front-page coverage of deforestation, the murder of Chico Mendes in Brazil, and attempts to arrange a debt-for-nature swap in Amazonia are all indications of this concern. In spite of this, however, many uncertainties remain.

Discussions of tropical deforestation are hampered by a lack of adequate data. As Myers (1984, 35) notes, "We do not have a precise idea of just how much tropical forest still exists—and how fast it is being cut down." Estimates of deforestation rates of tropical closed-canopy forests range from 78,000 km^2 (Lugo and Brown 1982) to 200,000 km^2 (Myers 1984) per year. This wide range of estimates at the global level is paralleled by an even wider range of estimates for individual countries (Allen and Barnes 1985). The range in estimates is caused by the fact that in many countries adequate statistics simply do not exist and, where statistics do exist, they may not be comparable because of the use of differing definitions of forest cover and deforestation.

As an example of the latter problem, Myers (1980) defines deforestation as the change from primary closed-canopy forest to any other use; the Food and Agricultural Organization/United Nations Environment Programme (FAO/UNEP) (1982) defines deforestation as the change of any type of closed forest to any other land use; and the FAO (1980) defines deforestation as the transformation of forestland where forestland is a more general category than primary closed forest or closed forest and includes land under agroforestry and shifting cultivation. Further, the issue of how to measure degradation of forest cover has not been resolved.

Even though the extent and rate of tropical deforestation are still open questions, most authors would agree that it is occurring in much of the tropical Third World (Allen and Barnes 1985) and many expect the rate to increase in the future as forest stocks decline and populations increase (Barney 1980; National Research Council 1982). Indeed, some authors feel that the rate of forest removal is already increasing (Bowonder 1985–86; Myers 1984; World Resource Institute 1990). At the same time, the difficulty of accurately determining national and global rates of deforestation should not obscure the fact that numerous authors have reported rapid rates of deforestation throughout the world: Brazil (Fearnside 1982), Africa (Anderson 1986), Nepal (Bajracharya 1983), Malawi (French 1986), Philippines (Revilla 1988), Haiti (Lewis and Coffey 1985), India (Nair 1985), and China (Primack 1988).

The differences in methodology and definitions used in determining rates of deforestation have been discussed by numerous authors (Grainger, 1980, 1983; FAO 1987; Lugo and Brown 1982; Melillo et al. 1985; Serna 1986); however, there is still no commonly agreed-upon definition of forest cover, forest degradation, or deforestation. As Allen and Barnes (1985, 169) note: "It is not clear . . . which data set [FAO *Production Yearbooks*; Lanly 1982; or Myers 1980] is most appropriate for analyzing the causes of deforestation." It should also be noted that in some cases there are no data sets altogether. While global monitoring of deforestation using remotely sensed data has been recommended, it is still not a reality (Grainger 1983; *Scientific American* 1986; Woodwell et al. 1983).

In this study, I define deforestation as "the complete removal of existing forests and their replacement by other forms of land use" (Palo et al. 1987, 55). This definition covers all natural forests but excludes trees on farms or woodlots. Deforestation is measurable, unlike forest degradation, which is the gradual decline in the production potential of a forest, something difficult to measure or quantify (FAO 1987). Degradation is usually the result of logging, grazing, or fuelwood collection. While degradation may be a serious problem in some areas, the concern of this research is with deforestation.

It should also be noted that the process of deforestation can occur either in one step, by clear-cutting, or in two steps, by selective logging followed by agriculture. It may be the case that there is a gap between forest degradation caused by logging and the final deforested state. This point will be discussed in more detail in chapters 4 and 5.

Since many observers have commented that tropical deforestation is primarily caused by "shifting cultivation" (also known as swidden or slash-and-burn agriculture), this term requires discussion. Following Lanly (1985), shifting cultivation is defined as a system of agriculture where crops are cultivated for a short period (usually one to three years) followed by a long period of fallow. It is possible to distinguish two broad categories of shifting cultivation, that practiced by traditional shifting cultivators and that practiced by migrants. In the Philippines, the Tagalog word *kaingin* is commonly equated with shifting cultivation; however, as will be discussed below, this equivalence of the two terms is inappropriate. As will be demonstrated in the chapters that follow, our lack of knowledge regarding tropical deforestation is matched by our lack of knowledge of what happens after deforestation. The operational difficulties involved in using a term as vague as "shifting cultivation" will be discussed in detail in chapters 3 and 4.

The harmful aspects of tropical deforestation have been discussed at length (Barney 1980; Caufield 1985; Eckholm 1979; Myers 1980, 1984; Plumwood and Routley 1982). Negative effects of deforestation can be separated into three broad scales: macro (global), meso (national and regional), and micro (local). The divisions are general but they provide a useful framework.

The macro effect of reduced tropical genetic diversity due to deforestation is probably the most important and has received the most attention in the literature (Barney 1980; Myers 1980, 1984; Sutlive et al. 1981a, 1981b). Among the concerns associated with tropical forest decline is the loss of potentially valuable tropical products as a result of extinction, the loss of genetic information as a result of the decline in the number of species, and the loss of the opportunity to study tropical ecosystems scientifically before they are destroyed or seriously disturbed.

A second macroscale concern expressed in the literature is the possible effect of deforestation on weather and climate patterns. The concern has two components. First, as the carbon stored in tropical vegetation is released as CO_2 either through combustion or decay following deforestation, CO_2 may increase in the atmosphere and contribute to global warming (Houghton and Woodwell 1989; Myers 1988b; Woodwell et al. 1983). Since the issue of atmospheric CO_2 and global warming is complex and cause and effect are difficult to establish, I recognize that a great deal of uncertainty surrounds the issue of tropical deforestation and its connection with global warming (Detwiler and Hall 1988; Myers 1988b). Second, deforestation results in changes in the micro-climate (ground temperature and albedo) and, possibly, rainfall patterns on a regional scale (Dickinson 1981; Myers 1988b).

At the mesoscale and microscale, the effects of deforestation which have received a great deal of professional and popular attention are large-scale soil erosion, land degradation, and flooding. The argument is that removal of forest cover (especially on sloping land) leads to soil erosion, increased runoff, and sedimentation that may increase downstream flooding during the rainy season. On the other hand, removal of forest cover may result in decreased stream flow in the dry season (Aiken and Moss 1975; Barney 1980; Bowonder 1985–86; Burgess 1973; Dasmann et al. 1979; Gomez-Pompa et al. 1972; Guppy 1984; Lewis and Coffey 1985; Myers 1980, 1984; Sioli 1985). These effects are particularly severe in the tropics owing to higher rainfall levels and greater intensity of rainfall.

The relationship between deforestation and downstream effects, such as flooding, is complex and controversial. Even such an apparently simple matter as measuring soil erosion is fraught with theoretical and practical difficulties (Blaikie 1985; Stocking 1987) and when there is evidence of increased flooding, its relationship to deforestation is unclear. In the case of the Upper Amazon Basin, Gentry and Lopez-Parodi (1980) and Nordin and Meade (1982) have argued as to whether or not there is evidence to support the claim that deforestation is causing increased flooding. In the case of the 1988 Bangladesh floods, Hamilton (1988) argues that there is no evidence to support the claim that deforestation in the Himalayas caused the floods.

In general, Hamilton (1982, 1985, 1986) has argued that removal of forest cover has only a minor effect on duration and impact of flood peaks and on dry-season flow in the tropics; however, a crucial aspect of his argument is that the previously forested land is allowed to return to its original cover. If the land is seriously degraded after or during deforestation, the argument may not hold.

Despite these reservations, most authors accept that land degradation is a serious and growing problem in the tropics (Blaikie 1985; Blaikie and Brookfield 1987) and that deforestation has resulted directly in soil erosion at the local level, resulting on a larger scale in increased sedimentation and runoff, which have contributed to downstream flooding (Myers 1984).

Additional mesoscale and microscale effects include the loss of homeland for peoples who live in tropical forests, which according to the World Bank (1978) may be as many as 200 million; loss of minor forest products such as rattan, resins, gums, and wildlife; loss of amenity and recreational resources; loss of fuelwood, which can assume regional dimensions, as the deforestation in the Sahel demonstrates; loss of timber resources and possible disruption of wood-based industries; and destruction of coral reefs as a result of sedimentation caused by erosion on deforested lands. The loss of fuelwood in the dry tropics may be the most immediate harmful effect of deforestation because in many tropical countries it is the largest source of timber drain (Bo-wonder 1985–86; FAO 1987). In Africa, deforestation has already meant large increases in time spent on fuel gathering (particularly by women) and a decrease in the number of hot meals cooked each day. In India and Nepal it has meant increased use of dung as fuel, which reduces the amount of fertilizer available for agriculture (Anderson 1986; Bajracharya 1983; Shrestha 1986).

Finally, we must recognize that tropical forests vary a great deal in terms of structure, composition, and environments in which they are found. For instance, Nair (1985) identifies sixteen types of forest for India, which range from mangrove swamps to tropical moist forests to dry, open forests that border on scrubland. Since tropical forests are not monolithic, rates, extent, and effects of forest use will depend upon forest type, intensity of use, climate, and geomorphology. Broad generalizations about tropical deforestation may be of limited use given the variability of forest types, physical environments, and socioeconomic or cultural activities affecting them.

Even though data on the effects of tropical deforestation are incomplete, it is safe to say that the process is occurring throughout the tropics in a variety of natural environments and under a variety of socioeconomic and cultural systems, that the rate of deforestation may increase in the future, and that the negative consequences are both immediate and long-term, local and global in scale.

The overall objective of this study is to analyze and model deforestation in the postwar Philippines at the provincial level. The goal is to further our understanding of deforestation in one country so as to better comprehend the complex nature of the process, increase our ability to choose policy interventions to control the process, and provide a case study that can be used to increase our understanding of tropical deforestation in general. The analysis will proceed in five steps:

1. The experiences of other tropical countries will be reviewed (particularly those of Southeast Asia) at the descriptive, theoretical, and quantitative levels to ascertain what factors or variables other writers have determined to be relevant for understanding tropical deforestation at the national, regional, and global levels.

2. The experience of the Philippines at the national and subnational levels regarding deforestation will be reviewed and the socioeconomic, demographic, and historical environments of the postwar Philippines will be analyzed to determine factors which appear to be relevant to understanding deforestation in the Philippines.

3. On the basis of steps (1) and (2) above, those variables most relevant to the Philippine situation will be chosen for analysis.

4. The variables chosen in step (3) will then be examined using multivariate analytical techniques.

5. The model constructed in steps (1) through (4) will be evaluated for its relevance to the Philippines and other countries.

Detailed case studies of tropical deforestation on a national basis are almost completely lacking (Grainger 1986) and little is known about the history of deforestation (Tucker and Richards 1983; Richards and Tucker 1988). Deforestation in temperate and tropical regions is not unique to the twentieth century, but the historical time paths of forest cover and land in agriculture are known only at very general levels. A number of questions relating to the change of forest cover through time would include the following: Is tropical deforestation more widespread geographically today than in the past? Is tropical deforestation occurring at a more rapid rate today than in the past? Answers to these questions may be essential for understanding the process of deforestation and predicting its future course.

Part of the difficulty is the sheer complexity of the problem, a complexity which exists within and between countries. Is each country a special case or are there generalities which pertain to all countries? Why does the rate of deforestation increase rapidly at times? While the literature would certainly seem to support the straightforward statement that "population pressure causes deforestation" or "population increases are related to deforestation," it is not as simple as that. First of all, the whole question of "What is population pressure?" is open to discussion. Is it a function of population density alone or are socioeconomic considerations also important? Second, even though population is obviously an important element in the deforestation issue, the precipitating factor may be the result of a government road-building program and not population pressure per se. Third, numerous societies in the postwar period have experienced increasing populations and increasing forest cover, such as Japan, the countries of Northern Europe, and the United States. Why is deforestation a problem mainly in less industrialized countries? Is rapid deforestation a stage that all countries go through, or is the deforestation in Third World countries in the twentieth century different, for example, from deforestation in nineteenth-century America?

Deforestation is a process which occurs on more than one level and there is no one single cause. One's interpretation of the multiple causes will depend upon the scale of reference used: at the local level the farmer or logger cuts down the forest; at the provincial level a new road is built which provides access for the two agents of destruction; and at the national level a logging concession is sold to a person with the right connections. Since each of these levels of analysis is relevant to an understanding of deforestation, the assignation of cause (or blame) is difficult.

The gaps in our knowledge regarding the rate and extent of tropical deforestation hamper analysis of the deforestation process and formulation of appropriate policy initiatives to manage tropical forests. This problem is compounded by a similar lack of knowledge regarding activities directly related to loss of forest cover; for example, relatively little is known about sustained-yield agricultural systems in the tropics (Janzen 1973) and tropical ecology (Proctor 1985).

Who Benefits from Tropical Deforestation?

The negative aspects of tropical deforestation are well known; however in light of the fact that deforestation continues, it may be appropriate to discuss those aspects of deforestation which are viewed in a positive light by certain participants in the process. The value of this approach is that it puts the emphasis on those persons who benefit from deforestation and therefore have a vested interest in its continuation. Some groups gain from deforestation and, from the viewpoint of policy intervention, the perspectives of these groups may be more important than those who do not. For any one country, it would be unusual for all of the groups discussed below to have a vested interest in deforestation; however, on a global basis, they appear to be the ones who are benefiting from forest destruction.

First, migrant farmers and traditional shifting cultivators benefit from the removal of forest cover because it permits them to farm for several years. It should be noted, however, that in most cases these benefits are short-lived, since the land clearance process must be repeated within two or three years. An exception to this would be traditional shifting cultivators, since swidden is integral to their culture and the benefits derived can be viewed as permanent (Conklin 1957). Although data on the relative numbers of traditional as opposed to migrant shifting cultivators do not exist for any country, the most likely case is that the latter greatly outnumber the former. Thus, the vast majority of people who practice short-term agriculture on previously forested lands do not derive any long-term benefit from it except for the subsistence and cash crops raised.

Second, a large number of different groups derive long-term benefits from tropical deforestation, including governments of countries with tropical forests, since pioneer settlement and the resulting deforestation divert attention from pressing social problems such as poverty and unemployment, particularly in urban areas; commercial loggers (legal and illegal) and those allied with them (politicians, military officers, and government bureaucrats); people employed in the logging and wood-processing industries; national treasuries that derive foreign-exchange earnings from forest products; commercial interests who utilize deforested lands to grow a product for the market; commercial interests who speculate on land near roads and new settlements; local commercial businesses who benefit from frontier settlement, such as banks and retailers; commercial interests who buy and sell charcoal or fuelwood; urban consumers of charcoal and fuelwood who pay a price which does not reflect the negative effects of wood harvesting and deforestation; transnational corporations involved in the forest product trade or the export of tropical products grown on previously forested land; and consumers in the First World, because the prices of tropical products do not incorporate the negative effects of deforestation. The groups enumerated above all benefit from deforestation in one way or another.

In short, it must be recognized that one of the most significant and immediate effects of deforestation is that it provides a stream of benefits to certain groups in society who therefore have a stake in its continuation. Since the primary purpose of better understanding tropical deforestation is to enable us to devise instruments for directing or controlling the process, an understanding of who benefits from the destruction of tropical forests is necessary to ensure that the instruments adopted are appropriate. The importance of who has benefited from deforestation in the Philippines will be discussed in chapter 5.

2
Tropical Deforestation: A Literature Review

Introduction

This chapter is concerned with the causes of deforestation as presented in the literature at the descriptive, theoretical, and quantitative levels. Given the extensive literature on the subject, a complete review is neither possible nor germane to this study. Rather, my purpose is to highlight what other authorities have considered to be the main causes of tropical deforestation.

The term "descriptive" requires an explanation; it is used to distinguish between those writings which are primarily theoretical, quantitative, or both and those which are not. The descriptive literature on deforestation, while it may enumerate what the author perceives as the causes of deforestation, makes no attempt to present this in a theoretical framework, test hypotheses, or statistically analyze data. This is not meant as criticism; rather, it is simply the recognition that much of the literature has not been empirical in nature or grounded in any theory of tropical deforestation. The disadvantage, of course, is that there is no way to determine which descriptive study is correct in terms of its ability to articulate the causes of deforestation.

While I shall review the experiences of countries in Latin America, Africa, and South Asia, relatively more attention will be paid to the countries of Southeast Asia (Indonesia, Malaysia, and Thailand). This is so for several reasons: (1) logging appears to have been a major factor in all Southeast Asian countries; (2) they are similar to the Philippines in terms of their culture and socioeconomic situation; (3) their vegetative cover is roughly similar to the Philippines; and (4) corrupt practices involving government officials, loggers, and the military seem to be the norm throughout the region.

This chapter is arranged primarily by geographic regions: descriptive and quantitative studies will be found mixed together rather than in separate sections in order to capture important different regional experiences with deforestation. The quantitative work will be separately analyzed in more detail in chapter 8 when it is compared with the results of this study.

Descriptive and Quantitative Work

Two recent edited volumes on deforestation in the nineteenth and twentieth centuries are Tucker and Richards (1983) and Richards and Tucker (1988) respectively. The contributions are primarily historical and they provide a good introduction to deforestation in general and tropical deforestation in particular.

In their volume on deforestation in the nineteenth century, Tucker and Richards (1983, xi) set forth the hypothesis that "in the course of the long nineteenth century before the cataclysmic global changes initiated in 1914, the steeply rising demand for the production of agricultural commodities exerted by the core or metropolitan societies of Europe, North America, and Japan was the dominant cause of rapid depletion of world wood and forest resources." For the tropical countries, this process involved several steps: demand from the core for tropical and semitropical agricultural commodities (such as rice, sugar, cotton, coffee, tea, and fibers) and increased cash crop production in the client country, which occurred primarily through an expansion of land under agriculture. This resulted in deforestation. Case studies include Brazil (sugar and coffee), Burma (rice), India (cotton), and the Philippines (coconuts, sugar, hemp). The editors note that plantation agriculture is an obvious cause of deforestation but that the secondary effects of plantations (the inability of peasant producers to feed themselves) may be even more important if poor people are forced to cut forests to meet their subsistence needs. Overall, they conclude that "the rate of change of tropical forest is as yet only guesswork for the years until the 1950s" (p. xvi).

In the second volume, Richards and Tucker (1988) note that global deforestation has continued in the twentieth century and, indeed, has accelerated in many cases; primarily in Third World countries. Once again, many of the forces leading to deforestation emanate from the developed countries through their demand for tropical products (timber and agricultural exports). Examples discussed include: India (tea), the Sahel (cotton and peanuts), Brazil (timber), and Thailand (rice, maize, cassava, kenaf, and timber). As in the nineteenth century, plantation economies producing for the world market resulted in impoverishment of the peasantry with resulting negative consequences for forests, a point also mentioned by Westoby (1981).

As an example of the importance of understanding the historical background of tropical deforestation, Lewis and Coffey (1985) note that most of Haiti's tropical rain forest was already destroyed by the end of the nineteenth century, and Eckholm (1976) demonstrates that in some parts of the world deforestation has been occurring for a long time, for instance, in Ethiopia, India, and Lebanon. In short, not all tropical deforestation is of recent origins; however, while the history of tropical forests is important in and of itself, the discussion that follows is mostly concerned with deforestation since 1945.

The Global Focus

Bowonder (1985–86) argues that the main causes of deforestation in developing countries are increases in cultivated land, cattle grazing, and fuelwood extraction due to population increases and poor forest management systems which have emphasized commercial exploitation at the expense of environmental conservation. In effect, Bowonder is arguing that population growth is the primary factor in deforestation.

In contrast, Plumwood and Routley (1982) argue that population increases and expansion of shifting cultivation are not the main factors behind tropical deforestation; rather, they place primary emphasis on socioeconomic causes. These include lack of employment opportunities for poor people, inequality in distribution of assets (particularly land), and fiscal incentives to certain groups who benefit from deforestation, such as cattle ranchers in Brazil, commercial loggers, and large-scale resettlement projects. The factors they discuss are very similar to Guppy's (1984) analysis, which will be presented below.

The World Resources Institute et al. (1985) argue that the single most important cause of deforestation is the expansion of agriculture. However, they are quick to point out that the ultimate causes of this agricultural expansion are socioeconomic in nature: poverty, low agricultural productivity, and the inequitable distribution of land. These factors are exacerbated by high population growth rates. They also note that governments in tropical countries have contributed to forest depletion by adopting policies which encourage the mining of forests and that developed countries are also partly responsible for this state of affairs because they provide a ready market for tropical forest products.

The role that governments play in deforestation has recently been analyzed by Repetto (1988, 1990) and Repetto and Gillis (1988). In an examination of ten countries and their policies toward the forestry sector, Repetto (1988) came to the conclusion that many governments have adopted misguided policies which have undervalued the forest resource and encouraged a more rapid depletion than is economically justified. Inappropriate policies include taxes which encourage the creation of inefficient wood-processing industries, subsidies to ranchers, and the granting of short-term concessions to commercial loggers. In short, forest resources have been sold at very cheap rates and the noncommercial services of the forests have been undervalued or ignored. Since governments are largely responsible for how forest resources are used, their policies have led to overexploitation in the name of development (Repetto 1988, 1990).

The World Bank (1978), in its first major policy statement on the forestry sector, cited two major reasons for deforestation: population growth and encroachment (primarily for food production) and uncontrolled commercial extraction. The underlying factors, however, are considered to be increasing population pressure, skewed distribution of land, lack of political will, and weak forestry institutions. Causes almost identical to these are reiterated by Eckholm (1979).

Myers (1980) argues that tropical moist rain forests are being converted at the rate of 245,000 km^2 a year. He attributes this to three major factors: logging followed by forest farmers (200,000 km^2 a year), fuelwood gathering (25,000 km^2 a year), and commercial ranching (20,000 km^2 a year and almost exclusively restricted to Latin America). Two caveats are in order regarding Myers's figures: first, they apply to moist forests only; second, they are, as Myers clearly states, "crude approximations." Myers's figures indicate that logging (legal and illegal) followed by migrant farmers and traditional shifting cultivators accounts for approximately 80% of all conversion of tropical moist forests.

Grainger (1980, 1986, 1987) has discussed tropical deforestation at length and his theoretical work will be reviewed later in this chapter. He considers population growth, increases in per capita GNP, and accessibility and environmental factors to be the three most significant determinants of tropical deforestation. These "mechanisms of deforestation" become apparent through the three major "types of forest exploitation": shifting cultivation, permanent agriculture, and logging.

Grainger (1986) attempts to verify empirically the mechanisms of deforestation by conducting a cross-sectional statistical analysis of forty-three countries in the humid tropics. His dependent variables are direct measures of deforestation (average annual deforestation rate, annual percentage national deforestation rate, average annual per capita deforestation rate); and his independent variables cover population and population growth rates, GNP per capita, average annual rates of growth of GNP per capita, total forest area, and several measures of accessibility such as river density and forest area logged. As a cause of deforestation, population growth could lead to removal of forest cover through an increase in cultivated land (shifting and permanent) and Grainger argues that this increase should be a lagged response to the growth of population. Cross-sectional correlation analyses determine a Pearson product moment correlation coefficient (r) of 0.56 between the log of the average annual deforestation rate and the log of the average population growth per annum from 1970 to 1980, which lead Grainger to conclude that population growth is the major force behind tropical deforestation.

As per capita income increases, demand for food and per capita food consumption should increase along with the intensity and areal extent of cultivated land. At the same time, the relationship between shifting and permanent cultivation is unclear, since future scenarios will depend upon population density, technical change in agriculture, and government programs. Grainger's results indicate that the average annual per capita deforestation rate rose with GNP per capita 1980 with a correlation coefficient of only 0.30. In addition, the annual per capita deforestation rate was correlated with the average increase in food production per hectare (1970–80) with a Pearson correlation coefficient of only 0.22.

Regarding accessibility and environmental factors, Grainger found the following:

1. Average annual deforestation rate was strongly correlated with forest area (r = 0.65).

2. Average percentage deforestation rate and river density were poorly correlated (r = 0.29).

3. Average annual deforestation rate was strongly correlated with area logged, 1946–79 (r = 0.60) and area logged, 1970–79 (r = 0.67).

4. Average annual percentage deforestation rate was poorly correlated with the percentage of national land area above 3,000 feet (r = 0.24).

Grainger then tested the following relationship for thirty countries using a stepwise regression with all variables in log form: deforestation rate = f (population growth, per capita income, accessible forest area). The results indicate that population growth and area logged (1970–79) were the two most significant explanatory variables. He then estimated the following equation, which is in log form. T-statistics are in parentheses below the appropriate variable.

Deforestation = –1.61 + 0.38 population increase + 0.49 area logged
$$(2.8) \qquad\qquad\qquad (2.7)$$
Adjusted $r^2 = 0.52$

Grainger concludes his discussion by noting that since deforestation mechanisms act in a spatial context, their relative significance should be determined by an examination of deforestation on a national or regional level. He also notes that no studies of this sort have been done.

Palo et al. (1987) conducted a cross-sectional analysis of forest cover in sixty tropical countries in 1980. The dependent variable was percentage forest cover (%FC) and the statistical analysis was conducted using correlations and multiple regression. The primary independent variables were: population growth, population density, gross domestic product per land area, GNP per capita, forest product exports per unit forest area, industrial roundwood production, food production per capita, agricultural area, fuelwood production, livestock production, and share of forest fallow (a measure of the extent of shifting cultivation). The authors tried various combinations of variables and groupings of countries, but owing to the importance of the population density variable and the relatively good data available, they decided to use only one equation for their alternative forest-cover scenarios for the 1980–2025 period. This equation is presented below with T-statistics in parentheses.

$$\%FC = 71.21 - 0.22 \text{ (population density)} \qquad r^2 = 0.60$$
$$\quad(27) \qquad\qquad\qquad (-9.5)$$

Palo et al. (1987) contend that increasing population pressure causes deforestation through the following mechanisms: increased fuelwood gathering, shortened fallow period for shifting cultivation, increased grazing, spread of agriculture and commercial logging, and infrastructure development. Increasing population pressure is seen as the driving force of tropical deforestation even if it is indirect in its effects. In contrast to Panayotou and Sungsuwan (1989) (to be discussed below), the authors do not include any price variables in their analysis and, indeed, feel that profits, prices, and costs play only a minor role in the deforestation process. Finally, the authors claim that their results agree with the findings of Grainger (1986) discussed above and Lugo et al. (1981), which will be discussed below.

The importance of the work of Palo et al. (1987) is highlighted by the fact that it forms the theoretical background to continuing FAO work on developing a model of tropical deforestation in their Forest Resources Assessment 1990 project. Scotti (1990), on the basis of Palo's work, assumes that a negative relationship exists between population density and percentage forest cover, and his statistical tests use a sample size of forty-seven subnational geographic units drawn from eight different countries (see Revilla 1991 for an interesting discussion of Scotti's work).

Another quantitative work regarding tropical deforestation on a global scale is by Allen and Barnes (1985), who conducted a cross-national study of deforestation among developing countries with a per capita income less than $3,000 and forest area greater than 5% of national land area. All data regarding forest cover were taken from the FAO *Production Yearbooks* and they showed an average annual decline of 0.7% for all countries, using 1968 as the base year. They used two samples: countries from Africa, Asia, and Latin America (n = 39) and countries from Africa and Asia (n = 25). The statistical techniques used were cross-sectional analysis for the years 1969 and 1978 and panel analysis over the 1969–79 period. Their individual variables fell into three broad categories: wood use, land use, and socioeconomic factors.

The socioeconomic variables include population and GNP, and Allen and Barnes (1985) hypothesized two relationships: population growth should be negatively correlated with forest cover, and GNP should be positively correlated with forest cover, because a higher GNP allows a switch to commercial fuels. It is interesting to note that while Grainger (1986) postulated a negative or positive relationship between GNP and forest cover as rising per capita income increased the demand for food and at the same time increased agricultural yields per hectare, Allen and Barnes (1985) suggested that the relationship would be positive. Land-use variables included arable land and land devoted to permanent crops, and the hypothesized relationship of the former with forest cover was positive and of the latter insignificant owing to the small percentage of permanent crops relative to land area. Wood use was measured by a composite index based on wood exports and fuelwood consumption. The expected relationship between wood use and deforestation was negative and, according to the authors, should occur with a lag because wood harvesting does not immediately destroy the ability of forests to grow again.

Allen and Barnes estimated two equations: equation (1) uses panel analysis (1969–78) and relates the annual change of forest cover to annual population growth, annual changes in cultivated land, annual per capita income growth, and annual changes in per capita wood use; equation (2), which attempts to capture the lagged effect of wood use mentioned above, uses a cross-sectional analysis to relate forest area (1968), population density (1968), per capita GNP (1968), per capita wood use (1968), and percentage area of plantation crops (1968) to changes in forest cover (1968–78).

For equation (1) the major results were that population growth was negatively related to forest cover for both samples with a level of significance of 0.10 and 0.05; the coefficient for change in arable land was negative (as expected), but the results were not statistically significant (however, because of the strong bivariate correlation between population growth and agricultural expansion, the authors concluded that changes in agricultural land were associated with deforestation); per capita wood use was not statistically significant in either sample and per capita GNP was not positively related to changes in forest cover.

For equation (2), the major results were that wood use in 1968 was negatively related to changes in forest cover from 1968 to 1978 for both samples and was significant at the 0.05 level; GNP and population density in 1968 were not significantly related to changes in forest cover; forest area was positively related to changes in forest cover for both samples at levels of significance of 0.10 and 0.05, which means that rates of deforestation were higher in countries with smaller forest areas; and permanent agriculture had a negative relationship to changes in forest cover in both samples at the 0.01 level of significance. Overall, Allen and Barnes concluded that growth of population and increases of land under cultivation cause deforestation in the short term and increases in wood use cause deforestation in the longer term.

Finally, Rudel (1989) conducted a cross-national study across thirty-six developing countries and found that deforestation was negatively related to population growth and the availability of capital.

A comparison of the results of the quantitative studies just reviewed indicates that, on a global scale, the factors that appear to be most important in explaining tropical deforestation are increasing populations and the spread of agriculture. Population in particular seems to stand out in this regard. However, at a deeper level, some authors feel that the real issue is access to the forest resource and social conditions which produce widespread poverty. In short, at the global level, it is difficult to articulate the specific causes of deforestation and it may be the case that such a general analysis is not very helpful in determining the causes of deforestation for a single country. The value of cross-national research in the study of tropical deforestation will be discussed in more detail in chapter 8.

Latin America

The area of Latin America which has received the most attention regarding deforestation has been Amazonia, primarily the Brazilian sector. The major factors behind the removal of forest cover in Amazonia have been identified as follows: large-scale government-sponsored and spontaneous settlement (Fearnside 1982, 1985; Smith 1976); road networks built to further migration and settlement (Fearnside 1986; *Scientific American* 1986); land speculation that allows the original settlers to sell their cleared holdings at high prices, providing them with the means to move on (Fearnside 1986); cattle ranching that follows pioneer settlement and is subsidized by the government (Binswanger 1989; Fearnside 1985; Myers 1980); military considerations behind some of the road and settlement projects (Smith 1976); and background forces such as population pressure and inequitable socioeconomic structures (Fearnside 1985; Myers 1984; Millikan 1988).

In a study conducted for the World Bank, Mahar (1989) argues that the major causes of deforestation in the Brazilian Amazon are small-scale agriculture, ranching, logging, road building, hydroelectric projects, mining, and urban growth. He also notes that the contribution of each of these to deforestation cannot be determined but that the major factor has been the expansion of agriculture. However, on a more fundamental level, he notes that poverty outside the Amazon region and government efforts to open up Amazonia to development have been the major forces behind deforestation.

In Central America, Williams (1986) argues that much of the deforestation has been caused by production of agricultural goods (bananas, coffee, cotton, and beef) for export and that this is a reflection of social inequalities such as the highly skewed distribution of income and assets, the extent of landlessness, and the appropriation of the best agricultural lands for export crops. Parsons (1976) also cites cattle ranching as a major factor in the conversion of tropical forests in Central America to pastureland, as does Myers (1984). Myers cites lack of land reform and inequitable social structures as major causes of settlement projects and landlessness.

Lugo et al. (1981) have analyzed deforestation in the Greater Caribbean area, which they define to include seven countries from Central America, four from South America and twenty from the Insular Caribbean, but not the United States. They note that most of the lowland forest was cleared before the end of the nineteenth century primarily as a result of monoculture for export (particularly sugar) but that since the 1940s deforestation has primarily occurred in those countries which have increased energy consumption to intensify land use and produce food.

Lugo et al. (1981) conducted a cross-sectional statistical analysis of deforestation in the Greater Caribbean area and their findings are that there is a significant inverse logarithmic relationship between population density and percentage of land forested (p = 0.001, r = 0.62); there is a significant relationship between the logarithm of energy consumption per unit of land area and percentage of land forested (p = 0.001, r = 0.45); when both population density and energy consumption per unit land area are regressed against percentage of land forested, r increases to 0.65; and there is a relationship between islands with steep topography (which they do not define) and percentage forest cover (r = 0.60). The authors' interpretation of this relationship is that higher elevations have more rainfall and therefore are less suitable for agriculture than areas found at lower elevations. In other words, moisture, rather than slope, may be the controlling factor. Another interpretation could be that forests at higher elevations have smaller trees and are more difficult to log. In any case, they claim that islands with steep topography have a higher percentage of forest cover than islands with little relief. In general, Lugo et al. (1981) place primary importance on population and land in agriculture, using energy consumption as a proxy for intensity of land use.

The only other quantitative study in Latin American is that of Reis and Margulis (1990). Their work in the Brazilian Amazon indicates that deforestation is positively related to population density, road density, and crop area.

Africa

As a contemporary phenomenon, deforestation in Africa is occurring primarily in the Sahel and sub-Saharan countries. While commercial logging is important in several countries in West Africa (Repetto 1988), it would appear that most deforestation is the result of expansion of agriculture and the cutting of trees for fuelwood (Anderson 1986; Gamser 1980).

One of the prime contributing factors to deforestation in Africa is that even though forests and woodlands are being depleted, fuelwood and charcoal are still cheaper than alternative fuels (Anderson 1986; French 1986). Referring to the work of Allen and Barnes (1985), Anderson (1986) points out that their hypothesized substitution effect of commercial fuels for fuelwood as per capita income increases seems to be weak in low-income countries. In addition, even if fuelwood were to increase in price, these price increases would make harvesting more feasible because they would increase the distance over which wood could be profitably transported and provide an incentive for poor, rural people to accelerate cutting. However, as work by Hosier and Milukas (1989) for Rwanda and Somalia indicates, charcoal prices in Africa rarely incorporate depletion effects, and charcoal prices in constant terms in both countries have not increased since the 1970s.

Whitlow (1980), in a study of Zimbabwe using aerial photographs from the 1960s and 1970s, argues that deforestation is part of a wider range of issues that include mismanagement of natural resources and rapid increases in human populations. Specific factors that have resulted in deforestation are the expansion of commercial and subsistence agriculture; the cutting of fuelwood for both rural and urban populations; the destruction of forests by elephants in game reserves, and uncontrolled fires. Whitlow (1980) states that expansion of agriculture, primarily because of population pressure in the tribal lands, has been the main factor in deforestation in Zimbabwe and, like Bajracharya (1983) for Nepal, he points out that fuelwood collection has become more critical because agricultural expansion is restricting collections to a smaller area of forest. In short, the main factors in deforestation in Zimbabwe have been increasing populations and the expansion of agriculture.

Hamilton's monograph (1984) is concerned with deforestation in Uganda in the twentieth century. Given the chaos that has increasingly come to dominate Uganda and the virtual breakdown of many government agencies, reliable statistics of any sort are hard to come by, but on the basis of historical research, surveys undertaken by students, and analysis of LANDSAT imagery, Hamilton suggests the following as causes of deforestation: fuelwood collection (wood represents 90% of Uganda's fuel consumption); expansion of agricultural land (it is interesting that since the major cash crops of cotton and coffee appear to have declined drastically in the 1970s and 1980s, most, if not all, of this expansion would be for subsistence agriculture); corruption within the forestry department and higher levels of government; and military security considerations.

In the case of Nigeria, Osemeobo (1988) points out that deforestation has been occurring for a long time: that is, forest cover declined from 60 million ha in 1876 to 9.4 million ha in 1985, a loss of 50 million ha in approximately 110 years. The main causes of destruction have been shifting cultivation, plantation agriculture, fuelwood gathering, and economic growth (i.e., industrial development, urban expansion, and infrastructure projects). Secondary causes include grazing, logging, and poor land-use management by the government. Forest depletion is also related to population increases.

South Asia

India is one of the few developing countries to have analyzed LAND-SAT images from separate dates to determine deforestation rates at the state and territory levels. The National Remote Sensing Agency (1983) reported that forest cover declined from 16.89% of total land area in 1972–75 to 14.10% in 1980–82, representing an approximate decline of 18%, or 2.6% a year, using the average of the beginning and end years as the denominator.

The already low level of forest cover in India is an indication that deforestation has been occurring for a long time. Richards et al. (1985) conducted a historical study of land-use change in parts of Bangladesh, Pakistan, and North India from 1870 to 1970, and their results show that forest cover throughout the region has declined during the entire period as agriculture and grazing spread and populations increased.

Bowonder (1982) notes that forest cover in India is rapidly decreasing and will continue to do so in the future. He also notes that, because of degradation, the quality of many of the remaining forests is declining. His enumeration of the causes of deforestation is lengthy and overlapping at times, but since this tendency to make lists of the causes of deforestation is common, we will discuss each in turn.

1. Ownership patterns: almost all forests in India are owned by the national government, but owing to government emphasis on commercial exploitation, the forestry sector has been characterized by low prices for wood and therefore low investment by commercial firms. In addition, corruption by government officials and logging firms has been widespread.

2. Fuelwood crisis: 90% of all wood harvested is used as fuelwood, primarily by low-income groups.

3. Encroachments: poverty and landlessness have resulted in the expansion of agriculture by shifting and permanent cultivators.

4. Housing needs: settlers, in search of housing sites, encroach on forestlands and cut down forests for housing material.

5. Transportation: new roads increase access to forest areas.

6. Agricultural operations: expansion of agriculture (food and plantation crops for export) results in the conversion of forestland. Bowonder (1987) claims that 70% of all land deforested between 1951 and 1976 was for agricultural purposes.

7. Irrigation and power projects: deforestation occurs through the submerging of forests and movement of workers to project sites.

8. Paper industry: increased consumption of paper has resulted in deforestation.

Nair (1985) shares Bowonder's concern about Indian forests and notes that most of the major problems facing forest managers in India are the result of societal complexities: changing power relationships among social classes, resource conflicts, and increasing population. Specific forces leading to deforestation have been agricultural expansion (caused by the lack of agricultural reform and the lack of good agricultural land) and the expansion of plantation crops (tea, coffee, rubber, sugarcane, and oil palm). In general, Nair argues that government forestry policy in India has been short-term in nature and conducted with an almost complete disregard for possible social and ecological consequences. In addition, the benefits of public-sector management have accrued primarily to the upper classes.

Nepal is considered by many to be one of the more dramatic examples of deforestation (Eckholm 1976; Myers 1984); although Blaikie (1985), Blaikie and Brookfield (1987), and Ives and Pitt (1988) question whether deforestation has been as great and the effect as damaging as presented in the literature.

Shrestha (1986) argues that there are four major causes of deforestation in Nepal: fodder collection, fuelwood extraction, timber harvesting (primarily for house building), and land clearing (primarily for subsistence agriculture). Bajracharya (1983), on the other hand, contends that the primary factor in deforestation has been the expansion of arable land because of chronic food deficits. He argues that while fodder and fuelwood collection contribute to deforestation, they become serious only after the amount of forestland has been considerably reduced by agricultural expansion.

Almost all of the deforestation reported for Nepal is being done by the poorer members of society, and the pressures motivating this type of activity can be viewed as a result of underdevelopment and a high population density. For example, 87% of Nepal's energy is derived from wood (World Bank 1978) and its person/cultivated land hectarage ratio is 6.5. An additional consideration is that in 1957 all forestlands were nationalized by the government, which accelerated deforestation because it took the control of forests out of local hands and, rather than have the lands taken away, people cut the forests and claimed them as private agricultural land. The same process was repeated in 1968 when landownership by cadastral survey became mandatory (Bajracharya 1983).

Southeast Asia

Of all tropical Third World countries, Indonesia, Malaysia, and Thailand are the most similar to the Philippines in terms of their overall socioeconomic structure and history of deforestation. Loss of forest cover in these countries has been rapid and widespread and particularly noticeable in the postwar period (Byron and Waugh 1988; Gillis 1988; Hamilton 1984; Meijer 1973; Panayotou 1983b).

One of the major characteristics these countries share (along with Brunei, Laos, Kampuchea, and Vietnam) is that the dominant family of timber trees in the lowlands is the same for all countries: the Dipterocarpaceae, more commonly referred to as dipterocarps (Asian Development Bank [ADB] 1987b; Weidelt and Banaag 1982). Dipterocarp forests are unusual for tropical forests because they have a large number of trees of the same species per hectare (for instance, in the Philippines, 60–90% of all growing stock comes from six dipterocarp species; Weidelt and Banaag 1982); subsequently, the yield (cubic meters of commercial wood per hectare) of dipterocarp forests is very high. In addition, the woods of the dipterocarps are highly prized by domestic and foreign users with the result that, compared to the rain forests of Africa or South America, the forests of Southeast Asia are of much greater economic importance (Weidelt and Banaag 1982). Like most tropical forests, the forests of Southeast Asia are predominantly broad-leaved and consist of hardwood trees (ADB 1987b).

As a result of their uniform stand structures and high yields, dipterocarp forests in all four countries have been heavily logged and a large part of the output has been exported (ADB 1987b). In the postwar period, approximately 70% of all tropical wood products have come from Southeast Asia (Callaham and Buckman 1981; International Bank for Rural Development 1972) and by the mid-1980s this had increased to 83% (Gillis 1988). The Philippines was the leading exporter until the mid-1960s, when it was replaced by Malaysia, which was in turn replaced by Indonesia in the 1970s.

Another characteristic common to all four countries has been corruption and illegal activity in the forestry sector. Details will be provided when I discuss each individual country, but the following evaluation by Callaham and Buckman (1981, 55) applies to the region as a whole: "Illegal practices occur on a large scale. Such practices are accepted as part of the heritage and ethic of the region. The problem seems to be more severe in some countries than in others. It is not likely to disappear or lessen, without major social revolution. The rich and powerful forces are likely to continue to prevail, and the current way of doing business will continue."

While corruption of some sort exists in all societies and corruption in the forestry sector occurs throughout the world, the immense profits to be made from exporting tropical timber has most likely led to more illegal activity in Southeast Asia than elsewhere. Although this statement cannot be proven or disproven, the summaries of the individual countries should make it convincing, and Repetto (1988) provides national-level data on the size of the economic rents in the forestry sector for Southeast Asia.

Climatically, Southeast Asia is characterized by high rainfall, often with an intense concentration in short periods of time (Aki and Berthelot 1974), monsoons (southwest and northeast), and stable temperatures (Duckham and Masefield 1969). Dry periods are common in some areas, such as northeast Thailand and southwest Philippines. On the basis of the Koppens classification, almost the entire region (with the exception of some highland areas) can be classified into Am (monsoon), Aw (dry winter), or Af (wet all year) (Mizukoshi 1971). It is difficult to generalize about soils since they are diverse, as is the geology (Duckham and Masefield 1969; Swan 1979).

The distinction between Af and Am/Aw climates is important because it is likely that deforestation has been more rapid in the Am/Aw climates (Whitmore 1984). The primary reason is that because of the longer dry season, Am/Aw vegetation is easier to clear and burn; in addition, it most likely burns at a higher temperature and thus ensures a more complete burn. This line of reasoning is consistent with the observation of Lugo et al. (1981) for the Greater Caribbean that there appears to be a positive relationship between forest cover and wetter climates.

Another common characteristic among the Southeast Asian countries is that forest area is almost entirely owned by the national government (with the exception of Malaysia, where it is owned by the state governments) and managed by government forestry departments (Panayotou 1983a). In addition, the major management objective of the forest departments has been commercial logging and the export of wood products (Byron and Waugh 1988), even though forests in all countries provide a gamut of products which directly affect the rural population, such as fuelwood, rattan, and bamboo. Forests in all countries have been the scene of conflicts among traditional occupants, commercial loggers, and migrants (Byron and Waugh 1988), and agriculture in all countries is diverse, with small-scale farming, plantations, and shifting cultivation all being important (Capistrano and Marten 1986; Conklin 1957; Panayotou 1983b; Spencer 1966). Agricultural yields, especially of lowland crops, have increased significantly since the 1970s (James 1983).

At the macroscale, the population, area (km^2), and population density (population per km^2) for these three countries and the Philippines are as follows: Indonesia (189.4 million; 1,904,000; 99), Malaysia (17.9 million; 329,000; 54), Thailand (55.7 million; 513,000; 109), and the Philippines (66.1 million; 300,000; 220). The Philippines has the greatest population density by far of these four countries. In addition, its population growth of 2.6% per annum is larger than that of Malaysia (2.5), Indonesia (1.8), and Thailand (1.5). In terms of GNP per capita (U.S.$), Malaysia is first with $1,870, followed by Thailand ($1,000), the Philippines ($630), and Indonesia ($430) (Population Reference Bureau 1990).

In terms of economic growth, percentage rates of increase in real gross domestic product for the period 1973–86 are as follows: Indonesia (6.6%), Malaysia (6.4%), Thailand (6.4%), and the Philippines (3.5) (ADB 1987b). In the 1980s, all four countries have had high rates of growth except the Philippines, which has done poorly. However, poverty, particularly among the bottom 20–40% of the populations of all four countries is still widespread, and serious regional and rural/urban disparities remain (Crone 1986; Hainsworth 1979).

Thailand. Thailand, on the mainland of Southeast Asia, is 80% larger than the Philippines with a lower population and a lower population growth rate. The population density of the Philippines is double that of Thailand. Thailand has had one of the most rapid deforestation rates in the postwar

Table 1.
Estimates of forest cover in Thailand since 1930

Year	Percentage	Information Source
1930	70	Thai Ministry of Commerce
1947	63	Thai Ministry of Agriculture
1959	58	Thai government estimate
1961	56	Estimate based on aerial photography
1963	53	Estimate based on FAO world forest inventory
1965	< 40	Estimate of forestry official
1969	52	Estimate based on aerial photography
1974	37	Estimate based on satellite imagery
1975	41	World Bank estimate based on satellite imagery
1978	25	Estimate based on satellite imagery
1980	< 30	Thai government agency estimate
1980	25.5	FAO/UNEP figure
1982	25	Estimate of Agrarian Land Reform official
1986	15	Unofficial estimate

Sources: Adapted from Feeny (1984) and Hirsch (1987).

period (Scott 1989; Sricharatchanya 1987) with forest cover declining from over 50% of land area in the early 1960s to approximately 20% in the mid-1980s (see table 1). Deforestation has been so rapid that by 1968 Thailand was a net importer of wood (Callaham and Buckman 1981).

Hirsch (1987) argues that deforestation in Thailand has been different in some ways from other parts of the Third World; in particular, land inequalities in Thailand are not as great as in other countries and the government has not encouraged large ranches or estates run by transnational corporations. In addition, Thailand is now a net importer of tropical timber, and fuelwood extraction is not based on subsistence needs. Finally, Hirsch argues that much of the observed deforestation cannot be attributed to shifting cultivation. Instead, he suggests that deforestation must be seen as the result of the development of the national economy over the past hundred years and cannot be blamed on any one single factor. However, he cites two factors as being of primary importance: the increase in area under cultivation and increased integration of the peripheral areas of Thailand with Bangkok.

The increase in area under cultivation has come about primarily at the expense of forested areas (particularly in the uplands) and has mostly been for commercial crops such as rice, kenaf, cassava, maize, and sugarcane. The rapid increase in roads documented by Hirsch was planned to support the expansion of commercial agriculture and for national security reasons. It has also facilitated logging operations. In short, Hirsch relates deforestation to the increasing integration of Thailand into the world economy, and to the needs of the Thai state to secure its boundaries and integrate peripheral areas of the country. He therefore views deforestation as part of the process of development and nation-building.

Feeny (1984) has written that the expansion of cultivated area has been the primary means by which Thailand has been able to increase agricultural production and exports, and that this has come about primarily through deforestation. Uhlig (1984) has described this process of frontier settlement and agricultural expansion in Thailand in detail, and Hutacharoen (1987) has analyzed deforestation from 1975 to 1985 in Chiang Mai province in northern Thailand. Hutacharoen's results indicate that conversion of forestlands to agriculture by lowland farmers and shifting cultivators was the primary cause of deforestation, with the impetus for this conversion coming from increasing populations and the efforts of the national government to increase exports of agricultural products. In the case of northeast Thailand, Nelson and Cruz (1985, 29) have pointed out that the area devoted to the production of cassava alone increased from 20,000 ha in 1970 to 674,000 ha in 1978, "all of which were previously under forest."

Sungsuwan (1985), in her M.A. thesis, attempts a quantitative assessment of those factors which have resulted in loss of forest cover in northeast Thailand. The study area includes sixteen provinces and forest-cover data are for four years : 1973, 1976, 1978, and 1982. Since the results of this work have subsequently been reported in Panayotou and Sungsuwan (1989), I will refer to this later publication for the remainder of this study. While aware that the open-access nature of the Thai forests is to blame for their destruction, Panayotou and Sungsuwan (1989) hypothesize that the following variables are related to loss of forest cover:

1. Population pressure, which increases demand for food, fuel, and forest products.

2. Lack of secure ownership, which reduces the incentive to make investments in land and makes agricultural credit more difficult to obtain, thereby increasing pressure on forest resources. (Note: since farmers may actually increase removal of forest cover in order to obtain title to land not now being farmed, secure ownership or the prospect of secure ownership may increase deforestation.)

3. Low yields of rice, which increase incentives for farmers to grow cash crops on newly cleared land.

4. Roads, which increase accessibility to forestlands.

5. The high price of wood, which increases incentive for logging (legal and illegal).

6. Lack of irrigation, which lowers opportunities to intensify cultivation and thus leads to forest clearing. (Note: if irrigated water is coming from a large waterworks, it may lead to deforestation through flooding.)

7. The high price of upland crops, which provides an incentive to clear more land.

8. The high price of kerosene, which encourages people to use fuel-wood and charcoal.

In order to increase the number of data points, Sungsuwan pools the cross-section and time-series data and uses a translog equation in an attempt to capture the interaction between the independent variables. However, since the interaction terms are not significant, the final estimated equations are log linear functions. The dependent variable is forest area as a percentage of provincial area and the independent variables and their expected signs are as follows: persons per square kilometer (–), titled area as a percentage of agricultural area (+), density of new roads (–), price per ton of cassava and maize (–), loans per unit of agricultural land (+), rice production per unit of land (+), price per cubic meter of wood (–), irrigated area as a percentage of agricultural area (+), price per ton of rice (+), price per liter of kerosene (–), and income per capita (+).

Since all of the independent variables are measured at the provincial level, it is assumed that interprovincial deforestation is not significant. Also important to note is that all forest-cover data are based on LANDSAT surveys and it would appear that the interpretation was done manually.

Owing to multicollinearity among crop, wood, and kerosene prices, Sungsuwan derives two models: one which includes the price of kerosene and one which includes crop and wood prices. Table 2 presents the results of the equation that does not include the price of kerosene. Panayotou and Sungsuwan (1989) note that all the coefficients have the expected sign and

Table 2.
Regression results of Panayotou and Sungsuwan (1989)

Variable	Regression coefficient	Standard deviation (Sx)	Standardized β coefficient ($\beta = \beta$ Sx/Sy)
Gross provincial product	0.42	0.48	0.29
Population density	– 1.51	0.30	– 0.65
Irrigated area	– 0.02	2.49	– 0.08
Relative price of upland crops and lowland rice	– 0.32	0.18	– 0.08
Price of wood	– 0.41	0.59	– 0.35
Rice yields	0.38	0.22	0.12
Distance to Bangkok	0.69	0.27	0.27
Density of roads	– 0.11	0.77	0.12

Source: Adapted from Panayotou and Sungsuwan (1989).
Notes: Coefficients have been normalized by calculating the β-coefficient so that the individual variables may be ranked in terms of their relative importance. All coefficients are significant at the 0.10 level with the exception of irrigated area, which is significant at the 0.15 level. Adjusted $r^2 = 0.77$; F = 32; df = 55; Sy = 0.69.

that population density is the most significant factor in explaining changes in forest cover. They suggest that population pressure works through three mechanisms: pressure to clear forestland for agriculture, demand for fuelwood and charcoal, and demand for construction material.

Of interest is that the variable for land security (titled land) is not significant. A possible explanation is that farmers actually cleared land in expectation that they would then receive title to the newly cleared land. Unfortunately, the data do not permit a test of this hypothesis; however, the importance of a proper land title has been given recent emphasis by Feder et al. (1988), who demonstrate that usufruct rights (government permits to farm on forestlands) do not increase access to credit. Land can be used as collateral for institutional investors only if there is a proper title. The relationship between land security and deforestation, therefore, remains unclear.

One factor in Thailand's deforestation mentioned by all of the authors in this section is illegal logging by people with political connections. The extent of this activity is difficult to assess but it appears to be widespread and serious (Callaham and Buckman 1981; Sricharatchanya 1987). Myers (1984) reports that approximately 55% of all wood processed in Thailand comes from illegal felling.

Based on this review of deforestation in postwar Thailand, the following factors appear to be relevant for understanding deforestation: population density, road networks, yield of agricultural land, irrigation, land security, logging, relative prices of food and commercial crops, relative prices of wood and alternative energy sources, and increases in agricultural land. It is interesting that while Panayotou and Sungsuwan (1989) consider agriculture important, they do not have a variable for agricultural land.

Malaysia. Deforestation has been increasing in Malaysia and in 1985 it was reported to be occurring at the rate of 250,000 ha per year (Repetto 1988). According to published reports, the three major causes of deforestation are: commercial logging, expansion of agriculture, and traditional shifting cultivation. It is also important to note that Malaysia is a fragmented country, with peninsular Malaysia a part of the Southeast Asian mainland, and the states of Sarawak and Sabah occupying the northern part of the island of Borneo.

The expansion of agriculture (primarily for rubber and oil palm) has been directed by the national government and much of it has come at the expense of virgin lowland forest and managed secondary forests (Myers 1984). Kwi et al. (1980) note that deforestation has occurred to promote development, especially agricultural development, and Ayub (1979), secretary-general of the Ministry of Agriculture, states that mines have precedence over agriculture and agriculture over forests. In peninsular Malaysia, Repetto (1988) notes that conversion of forests to agriculture has accounted for 90% of all deforestation in the past decade, with logging playing a significant role from 1955 to 1980.

The deforestation in peninsular Malaysia is one of the few examples in the tropical Third World where the original forest cover has been replaced in the past thirty years or so with a system of agriculture which is permanent and successful (Gillis 1988). Malaysia is now the world's largest exporter of rubber and oil palm and the agricultural sector has been responsible for providing jobs and improving per capita incomes (Repetto 1988; Scott 1989). While such large-scale clearing of forests is economically successful, concern has been raised about the environmental effects (Aiken and Moss 1975; Chuan 1982; Daniel and Kulasingam 1974).

Commercial logging, which began on a large scale in the late 1950s after the communist insurgency was put down (Burgess 1973), is important to the Malaysian economy in terms of employment and foreign-exchange earnings and has resulted in loss of forest cover (Repetto 1988). Byron and Waugh (1988, 56) point out that "the average area logged commercially was 400,000 ha per year over the period 1970–1986." This represents 6.4 million ha or approximately 19% of the total land area of Malaysia. According to Repetto (1988), in Sabah shifting cultivation has caused a little over 50% of all deforestation with the remainder accounted for by logging followed by agriculture, and in Sarawak the two major causes are logging and shifting cultivation. Detailed analyses of logging in the state of Sarawak and its effect on native groups and the environment are provided by Hong (1987) and the Institute of Social Analysis (1989).

In short, the three major causes of deforestation in Malaysia are commercial logging, expansion of agriculture, and shifting cultivation (primarily in Borneo). The shifting cultivation referred to here is the traditional variety: spontaneous migration of lowland farmers to forest areas does not seem to be taking place on a large scale. In addition, removal of forest cover for fuelwood does not seem to be a major problem.

An element of logging in Malaysia which can not be proven with any degree of certainty is corruption. The use of logging concessions as political favors has been pointed out by *Asiaweek* (1988), Kumar (1986), and Scott (1989). *Asiaweek* (1988) notes that the environmental minister in Sarawak controls several concessions, as do important politicians in the state. In the most comprehensive study of corruption in tropical forestry that I am aware of, Tharan (1976) has documented large-scale financial abuses in the state of Kelantan in peninsular Malaysia for the period 1960–74.

Indonesia. The forest area of Indonesia is almost three times that of Thailand, Malaysia, and the Philippines combined (ADB 1987b). In fact, it has the largest area of rain forest of any country except Brazil (Myers 1984). Commercial logging really did not start until after the fall of former president Sukarno in 1965–66 and the period since has witnessed rapid growth in the number and extent of concessions and exports of wood products (Repetto 1988). Only in the last ten years or so has the government started to exercise control of the industry and widen its view of forestry to include social and ecological considerations (Callaham and Buckman 1981; Soerianegara 1982).

Indonesia has moved rapidly from exporting logs to exporting plywood and now supplies 70% of the world's tropical plywood. Log exports have declined from 17.6 million m^3 between 1976 and 1980 to 0.3 million m^3 in 1985 (Byron and Waugh 1988). Even though logs are rarely exported in their raw form, approximately 50% of all remaining Indonesian forests are under logging concessions and deforestation is occurring at approximately 700,000 ha per year (Repetto 1988). The major cause appears to be expansion of agriculture, with commercial logging a contributing factor (Repetto 1988).

A conspicuous feature of Indonesian agriculture is the role that shifting cultivation plays in providing commercial crops for domestic and foreign markets (Dove 1983). Pelzer (1978) points out that in 1960, pepper, benzoin, coffee, copra, and rubber provided 40% of the total value of Indonesian non-oil exports and the majority of these were grown by shifting cultivators in Sumatra, Kalimantan, and Sulawesi. Roche (1988) notes that perennial trees (cloves, fruits, and coffee) and cash crops (sugarcane and vegetables) are increasing in importance in the uplands of Java. He also argues that the spread of cash crops and the integration of the uplands into the Indonesian national economy are the solutions to upland degradation and soil erosion. It should be noted, however, that the generally rich soils of Java make this more likely than in Borneo or peninsular Malaysia, where the soils are not as fertile.

The commercialization of agriculture in the upland areas of Indonesia is very similar to what has happened in Thailand with the rapid spread of crops such as cassava and maize and is to be contrasted with agriculture in Malaysia, where almost all commercial crops are in the lowlands, most often associated with government settlement projects. The situation in the Philippines is unclear, but it appears that the large-scale commercialization of upland agriculture has not occurred. An exception to the above characterization would be the growing of temperate crops near Mount Kinabalu in Malaysia and near Baguio City in northern Luzon.

A point of introduction to the issue of corruption is the East Kalimantan forest fire in 1982–83, which burned approximately 3.5 million ha (Mackie 1984). While the drought associated with El Niño was undoubtedly a leading factor in the blaze, logging waste and logging roads greatly facilitated the fire (Mackie 1984). It appears that much of the logging was illegal and linked to leading Indonesian military personnel (personal communication, anonymous sources). Plumwood and Routley (1982) have drawn a direct link between the military and political elite and forest concessions, and Repetto (1988) also alleges that corruption is widespread. In fact, for the 1979–82 period alone, Repetto (1988) argues that of the potential U.S. $4.4 billion in forest revenues, only U.S. $1.6 billion was actually collected, with the remaining U.S. $2.8 billion going to excess profits and payoffs.

The only quantitative work on deforestation in Indonesia is by Gastellu-Etchegorry and Sinulingga (1988), who designed a geographic information system to study deforestation from 1946 to 1981 in a 27,000 ha forest in central Java. They included fourteen parameters: soil type, elevation, slope, land erosion, pluviometry, population density, population increase, per capita income, number of farms, average land area owned by a family, geomorphology, watersheds, administrative boundaries, and forest cover. In declining order of importance, they concluded that the need for arable land, fuelwood, and fodder were responsible for deforestation, and that the highest deforestation rates occurred on lands characterized by good soils, lower slopes, higher elevations, middle population densities, high population increases, high percentages of farmers, large fields, and low incomes. The only statistic used was the coefficient of correlation and, since no information regarding statistical significance was presented, it is difficult to evaluate their results.

Summary

The literature just reviewed is difficult to summarize for several reasons. First, since the authors are not using an agreed-upon set of definitions, many of the suggested causes of deforestation may overlap or simply be incompatible. Second, in most cases, no indication of the relative importance of the various causes is presented, so that, for example, population growth and increased paper consumption are both listed as causing deforestation although population growth would certainly be a more important factor. Third, few authors discuss the links between the causes (e.g., population growth leads to expansion of agriculture, which then results in deforestation); therefore, it is difficult to distinguish between primary and secondary factors. Fourth, the regional and global studies are not directly comparable because the greater specificity of the regional research allows for the inclusion of more variables. Thus, a neat summary using clear-cut definitions of the causes of deforestation is not possible.

Based on their frequency in the literature, the importance attributed to them by their respective authors, and the quantitative work presented on their behalf, the following would appear to be the major causes of deforestation on a global scale: increases in population, inequitable social structures, expansion of agriculture (food and commercial crops, grazing), fuelwood gathering, logging, and corruption or mismanagement. The quantitative work alone identified the following as significant causes at the global level: increases in population (Allen and Barnes 1985; Grainger 1986; Palo et al. 1987; Rudel 1989; Scotti 1990), wood harvesting (Allen and Barnes 1985; Grainger 1986), and expansion of agriculture (Allen and Barnes 1985).

On a regional basis, the following factors were important: in Latin America, cattle ranching, resettlement and spontaneous migration, expansion of agriculture, road networks, population pressure, and inequitable social structures; in Africa, fuelwood collection, logging, expansion of agriculture, and population pressure; in South Asia, population pressure, expansion of agriculture, corruption, fodder collection, and fuelwood gathering; in Southeast Asia, corruption, expansion of agriculture, logging, and population pressure. The quantitative work of both Lugo et al. (1981) for the Greater Caribbean and Panayotou and Sungsuwan (1989) for Thailand indicated that population was the most important explanatory factor.

Comparing across regions, the following generalizations regarding the causes of deforestation appear warranted:

1. Cattle grazing is more important as a cause of deforestation in Latin America than elsewhere.

2. Fuelwood gathering seems to be a much more severe problem in Africa and South Asia and is at least partly related to the drier climate of these areas.

3. Corruption in the forestry sector seems to be a more significant factor in explaining deforestation in South and Southeast Asia than in Africa or Latin America.

4. Logging is or has been important in Western Africa and Southeast Asia.

5. Large-scale movement of people (government-sponsored and spontaneous) seems to be of primary importance in Amazonia.

At the very least, this survey indicates that tropical deforestation is not a monolithic process; not only is there a large variety of causes of deforestation, but these causes also operate in a wide variety of environments. At the same time, virtually all researchers agree that the single most important cause is population growth. While this summary is useful in indicating the range of causes and, to a certain extent, the relative importance of some of them, it does not mean that we have a theoretical framework in which to analyze the process of deforestation.

Theory

Theoretical work on tropical deforestation is not extensive. Most work has been descriptive and very few authors have attempted to place the process or processes of tropical deforestation into a larger framework. The lack of data and the dearth of quantitative and theoretical work on tropical deforestation are most likely interrelated: lack of data makes it difficult to formulate and test hypotheses and lack of theory makes it difficult to formulate guidelines for empirical research.

Of the theories reviewed below, three authors deal explicitly with tropical deforestation: Grainger, Guppy, and Walker. The work of Blaikie and Brookfield is discussed because they are concerned with land degradation in Third World countries and deforestation is part of this process. The remaining theory (current extensions of disaster theory) is included because it is directly relevant to tropical deforestation.

Grainger's Work

Grainger (1986, 1987) has written extensively about modeling tropical deforestation, although Poore (1976) was most likely the first researcher to present a theoretical analysis of the effect of human activity on tropical forests (Grainger 1989). Central to Grainger's thesis is the distinction between "types of forest exploitation" and "mechanisms of deforestation." Types of forest exploitation are forms of land use which replace forest cover. The three main types of forest exploitation are shifting cultivation (traditional, short-rotation, encroaching, and pastoralism), permanent agriculture (permanent field crops, resettlement schemes, ranches, plantations, settlements, and urban areas), and logging. Deforestation is the end result of all three types of forest exploitation; however, the types of forest exploitation are essentially reflections of the entire socioeconomic framework.

The pressures emanating from society as a whole which are the "primary causative factors" of deforestation are called "mechanisms of deforestation." Grainger suggests the following as of primary importance: population growth, economic development (increase in per capita income), and accessibility and environmental consideration. According to Grainger, the rate of increase or decrease of the types of forest exploitation, and therefore of the rate of deforestation, is determined by the mechanisms of deforestation. His tests for these factors on a cross-national basis were presented above.

Since Grainger's main concern is to understand how tropical deforestation actually occurs, he constructs a systems model for national land use which incorporates both mechanisms of deforestation and types of forest exploitation. The key element in this model is that conversion of forest to nonforest uses is the major process connecting the agricultural and forestry sectors; the driving forces are increases in population and per capita income, which increase demand for food and wood. The increase in food supply can result from an increase of area cultivated or increased productivity and it can be a result of shifting or permanent cultivation. An important feature of Grainger's model is his emphasis on the distinctive characteristics of land; the forest/agriculture interaction will be different on soils of high fertility and high sustainability than on soils of low fertility and low sustainability.

Within the forest and agriculture sectors are positive and negative feedback loops; that is, forces which tend to accentuate or counteract the original activity, respectively. In forestry, supply and demand for forest products and price are the most important variables; however, accessibility of forests, time and effort spent on forest protection, efficiency of utilization, and economic incentives to invest in forestry plantations are also relevant considerations. If old-growth forests are extensive (as in the Philippines in the 1950s and 1960s and in Malaysia and Indonesia today), negative feedbacks (increase in wood prices, need for forest protection, increased efficiency of utilization, and incentives to invest in plantations) may be weak or nonexistent.

In agriculture, per capita food consumption should increase as incomes rise, and this can increase the area or intensity of cultivation for both shifting and permanent agriculture. Thus, the interactions are dependent upon the relationship between shifting and permanent cultivation, prices for outputs and inputs, length of fallow period, closeness to markets, and quality of land. A potential negative feedback is degradation of agricultural land due to lack of investment in reducing soil erosion (a point also made by Fearnside 1986 and Blaikie 1985).

Grainger (1987) states that his model needs more research to be used reliably. Its main purpose is to make explicit the complex nature of the deforestation process and facilitate intervention by governments concerned with its adverse consequences.

Grainger's system model includes external demand for tropical timber and agricultural products; however, the relationship between the supply and demand for timber and price for timber and possible negative feedbacks on deforestation may not be as clear as postulated by Grainger. In the case of the Philippines, the forestry sector has been primarily export oriented. If foreign demand is essentially inelastic or if commercial logging (legal and illegal) is not price-sensitive, either because profits are so great or because the desire to acquire and hold dollars is so strong, then it may be the case that prices in foreign markets do not affect domestic production that much.

Guppy's Work

Guppy (1984) sees deforestation occurring as a result of four separate but related factors: (1) rapid population growth and land hunger, which are the ostensible causes of deforestation; (2) inequitable social conditions, which mean that landownership is highly skewed in favor of the elites and that little land, particularly good agricultural land, is available to the majority of people; (3) the political motivations of the local elites, which are primarily to enrich themselves and maintain power; and (4) the active support that governments of less developed countries (many of them with tropical forests) have received from Western countries and international lending agencies through increased funding, which has enabled them to remain in power even though most of them have been nondemocratic.

Guppy paints a scenario where foreign loans support and strengthen the governments of less developed countries, which become increasingly centralized and infatuated with large-scale development projects that foster forest destruction, such as dams, roads, and colonization projects. As the local elite grows, it becomes increasingly isolated from the mainstream of the country and corruption increases. Imported technology, increasingly tied to foreign loans and development projects, creates a new elite of technocrats closely allied with the military. Agriculture and land reform are neglected and an export-oriented development strategy is required to pay back the foreign loans. The distribution of income and assets worsens and poor people are driven to occupy the forests because of landlessness or oppression. As dissidence grows, so does the military.

The verbal model that Guppy presents clearly places more emphasis on political factors (especially elite control) than Grainger does. Guppy argues that much forest destruction (either directly through logging or indirectly through planned colonization) is a conscious attempt by local elites either to directly enrich themselves or to avoid dealing with problems such as poverty and landlessness. In this view, the forests are the means to placate social discontent and provide an income stream to those who control access to their use. A corollary to this hypothesis is that the land shortage is more apparent than real; instead, the real issues are land distribution and access.

Another difference between Guppy and Grainger is that the former appears to place more emphasis on influences emanating from outside national boundaries as factors important to understanding tropical deforestation. These influences are twofold: first, the support which more developed nations have provided to prop up many governments in the Third World and, second, the market which these nations have provided for tropical forest products. For Guppy, tropical deforestation is intimately related to the global, capitalist system. While Grainger includes overseas demand for logs, he excludes foreign demand for agricultural products. It may simply be a question of emphasis, but Guppy seems to stress the international context in which tropical deforestation is occurring more than Grainger does.

Walker's Work

Walker (1985, 1987) develops two models of deforestation using optimization theory. The first model and its two variants posit a commercial logger who may or may not abandon the concession to an agriculturist after completion of logging. The second model posits a landowner who must decide whether or not to use land for forest preservation or agriculture (or an alternative land use which yields a net economic benefit).

Model #1. Walker notes that multiple plots in a large forest can be optimized if the logger actually owns the forest; in this case, the logger is confronted with an infinite income stream because "the life cycle of [the] trees is small compared to the planning horizon" (1985, 3). However, in many developing countries, concessions of finite duration are granted, and therefore concessionaires will replant and exclude others from using the land only if it is profitable for them to do so. If not, logged land will be abandoned.

In Walker's "time-constrained case," the area granted to the concession is larger than the area that can be harvested during the life of the concession;

thus, there is no incentive for the concessionaire to replant and adopt exclusionary practices. In his "land-constrained case," the life of the concession is longer than the time needed to harvest the area granted to it. Walker derives the decision rule that the logger will replant and exclude others if the discounted value of harvesting the planted trees is greater than the cost of replanting and if the growing period of the planted trees is no greater than one-half the length of the concession. Thus, for the logger to engage in exclusionary behavior for the entire concession, the concession must be very long-term or the newly planted species must be very fast growing.

It is at this point that one of the limitations of Walker's model becomes obvious: he does not specify what type of trees are to be replanted. If the trees are of the dipterocarp type (as in Southeast Asia), there are two problems: first, the technology for creating dipterocarp plantations does not exist, and second, even if the technology existed, the time to harvest would be much longer than any concession. If the trees are fast-growing species, they are not substitutes for the trees originally logged and they would most likely be profitable only if the concession had a pulp and paper mill. In actuality, the real decision is not to replant and exclude others but to practice selective logging in a sound manner and exclude others so that a second cut is possible.

Model #2. Given competing uses for forestland, the owner will attempt to maximize the net present value of the two or more activities. The allocation of land will depend upon the competing returns and the discount rate. It is conceivable that forest cover will be reduced to zero in this formulation, since the positive externalities of forest cover yield no return to the owner and the negative externalities are not considered as costs.

The value of Walker's work is that it highlights the importance of the institutional rules that concessionaires are supposed to operate under; in particular, the number of years and area of the concession. Others include: forest charges, the maximum annual allowable cut (AAC), selective logging, rules regarding replanting, and guidelines for logging road construction. Since concessions, at least on paper, are not given carte blanche, all rules and penalties for not following the rules are essential for a complete understanding of concessionaire behavior (including deforestation).

A criticism of Walker's model (and of most forestry optimization models) is that forests provide multiple benefits of which timber is only one; others would include amenity, recreational, environmental, and, possibly, climatic benefits. In the Third World, minor forest products such as gums, oils, rattan, bamboo, and wildlife can be very important for rural people. None of these other uses are presented, and, in fact, it is not clear that they could be modeled in a meaningful manner. One use of tropical forests which is almost entirely ignored is that millions of people live in them.

The discussion by Walker can be extended to include the possibility of overexploitation or severe depletion as presented by Clark (1973). Clark argues that extinction of a renewable resource (in his example, fisheries) is possible under private management where the goal is the maximization of net present value. The key is the rate of discount used, since a high discount rate can lead to extinction of fish populations.

Since the rate of discount is so crucial to optimization models, it may be appropriate at this point to talk about discount rates in the Philippine context even if it anticipates some of the discussion which will appear later. Annual interest rates vary a great deal in the Philippines from 15–20% for government notes to approximately 1,000% for people who borrow at 6/5 for a week (6 pesos must be returned for 5 pesos borrowed the week before). The extreme case would be street vendors who borrow 6/5 on a daily basis. As will be discussed below, two of the main agents of forest destruction in the Philippines have been loggers and poor migrants who grow subsistence crops: in both cases, they have very high subjective rates of interest.

If subjective (time preference) and objective (actual) rates of interest are high and the common property resource is not protected, then a logical extension of optimization models is that the resource in question becomes a fleeting resource: that is, a resource which one can capture only by use, before someone else does. This argument will be developed more fully when we discuss the Philippines. In spite of the above criticisms of optimization models, the value of Walker's model is that it brings to the fore the rules of the game under which the participants of forest destruction must operate.

Blaikie and Brookfield's Work

Blaikie (1983, 1985) and Blaikie and Brookfield (1987) have written extensively about soil erosion and land degradation in Third World countries. They do not attempt to model deforestation per se; however, tropical deforestation is an important component of tropical land use. In this way, their analysis, with slight modification, is applicable to tropical deforestation, and what follows is an amended version of the model they have developed to analyze and explain tropical land degradation.

Blaikie and Brookfield note that, because of the multiplicity of environments, national and regional histories, cultures, and socioeconomic forces operating in the Third World today, no single or unicausal model of land degradation is possible. Rather, an explanation, especially one which claims international validity, must be able to encompass a wide variety of circumstances. The main question posed by Blaikie and Brookfield is: Why is it that land-saving or land-improving investments are not being made in large areas of the world?

The focus of their analysis is the land manager. This is the person or persons who make the immediate decisions regarding land use, including, for instance, what crops to grow, what inputs to use, and whether or not to make investments in land such as terracing or contouring. A crucial issue for the land manager is his or her access to resources, whether they be land, credit, or technological inputs. The environment in which the land manager makes these decisions is complex and includes ecological and socioeconomic variables. The land manager is nested in a series of relationships which move outward from the family or group to other groups in the area and the state, and to foreign influences. An analysis of why the land manager does or does not make investments in land will thus depend on the scale of the investigation.

Blaikie and Brookfield call this approach "regional political ecology": regional because it deals with the areal variation in land resources, political because it deals with the political economy of land-use decisions, and ecology because it deals with the interrelationships between the environmental and socioeconomic spheres and among the different scales from family on up to the global perspective. Politically, Blaikie and Brookfield take for granted that different classes exist and have different interests. In particular, they argue that the state is not neutral (it is most responsive to the interests of the elites) and that, most of the time, it has adopted an extractive approach to natural resources.

The global scale is important primarily in considering the foreign aid given to poorer countries, the markets that developed countries provide for resources, and the transnational corporations that are actively involved in agriculture and resource extraction in the Third World. The interface between the external sector and the state is particularly important because, since the state represents elite interests more effectively than nonelite interests, it can have a major say in what types of foreign activity will be allowed. Since environmental degradation and increasing rural poverty do not directly affect elite interests, the elite tend to show little concern for conservation or investment in rural areas.

One of the major virtues of the Blaikie and Brookfield exposition is that the land manager, the one who actually makes the land-use decision, is placed at the center of the analysis. In the context of the Philippines, the major users of the forests have been loggers (legal and illegal) and agriculturists (traditional and migrant). The methodology of Blaikie and Brookfield allows the context in which these land managers operate to be made clear.

Disaster Theory

Tropical deforestation and the inappropriate land use which follows have been described by many as an ecological/environmental disaster (Myers [1980, 1984] has been one of the strongest and most consistent commentators in this regard). The tragedy is that deforestation is destroying a resource with multiple uses and replacing it, in most cases, with an extremely unproductive agricultural-economic system. Blaikie and Brookfield and Guppy have placed socioeconomic and political elements at the center of their analysis of this process; a contemporary critique of "natural disasters" by several authors carries this analysis one step further.

According to Hewitt (1983), natural disasters have traditionally been viewed as a function of nature; they are unforeseen, extraordinary; and human activities are not considered important in the determination of what is or is not a disaster. Concomitant with this view of disasters is the belief that everyday life is foreseen, ordinary, and in general stable and predictable.

In contradistinction to the above view, Hewitt (1983) suggests that disasters reflect the relationship between society and nature. Within this framework, a disaster can be defined *"as the interface between an extreme physical event and a vulnerable human population"* (Susman et al. 1983, 262; italics in the original). The implication is that society, to a large extent, creates groups of people who are vulnerable to extreme physical events. The investigation of why or how society places people in vulnerable situations then becomes the focus of study.

According to Watts (1983), people, particularly poor people, are constrained by nature and the socioeconomic relations within society. These constraints are particularly widespread and confining in underdeveloped nations which are being integrated into the global, capitalist economy (Susman et al. 1983; Watts 1983). During this transition period, the traditional moral economy of the peasantry may break down and elite groups (domestic and foreign) may be able to gain control over productive assets, including natural resources. As an example, Watts (1983) documents how the "margin of security" for the Hausa peasantry of Nigeria declined in the twentieth century as a result of colonial and capitalist penetration.

For the underdeveloped countries as a whole, Susman et al. (1983, 266) argue that their populations are becoming increasingly vulnerable to physical extremes, and they link this directly to "the continuing process of underdevelopment." People are becoming marginalized because they are unemployed, without land, or on land which is marginal in terms of productivity and sustainability. Poor people are forced to migrate to areas of high geophysical vulnerability and engage in practices which only increase this vulnerability. "Ongoing underdevelopment is placing marginal people in marginal lands" (p. 280).

For our purposes, the tragedy of deforestation and inappropriate land use in the tropics can be viewed as a reflection of "the structure of social systems" (Watts 1983, 258). The recent work in disaster theory reinforces the analysis of Guppy and Blaikie and Brookfield regarding elite control and the lack of access to resources on the part of poor people. It places primary importance on the notion of vulnerability, mostly on the part of poor people.

3

The Decline of Philippine Forest Cover

Introduction

The Philippines at present has seventy-three provinces divided into twelve regions plus the National Capital Region (also referred to as the Metro Manila Area). In addition, the Philippines is often divided into four superregions. The names of these superregions and the main islands and regions included in them are as follows: Luzon (Luzon, Mindoro, Masbate; regions 1, 2, 3, 5, and part of region 4), Visayas (Panay, Negros, Cebu, Bohol, Leyte, Samar; regions 6, 7, and 8), Mindanao (Mindanao; regions 9, 10, 11, and 12), and Palawan (Palawan; part of region 4). The political and administrative divisions of the Philippines and percentage forest cover in 1980 are shown in figure 1.

The government agency in charge of Philippine forests has changed its name three times in the postwar period. The Bureau of Forestry (BF) became the Bureau of Forest Development (BFD) in 1973, which in turn became the Forest Management Bureau (FMB) in 1987. Since data from each of these agencies will be presented, all three abbreviations will be used. However, this should not obscure the fact that all three agencies are basically the same. On a yearly basis since 1968, these agencies have published *Philippine Forestry Statistics (PFS)*, which is the official source of forestry data in the Philippines. Since *PFS* has been published by three different agencies, all references will be to *PFS* and year of publication rather than to the agency under which it was published.

The major source of forest-cover data in this study comes from the recently completed Philippine-German Forestry Resource Inventory Project (1988) (henceforth abbreviated P-GFI). The results of the P-GFI have been published by the FMB; however, for ease of reference, all citations will be to P-GFI

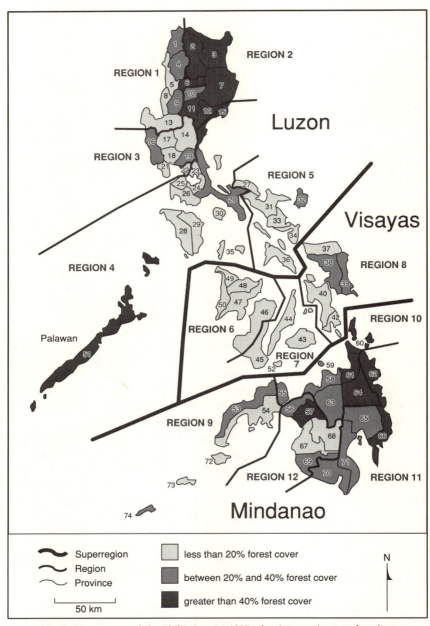

REGION 1

REGION 2

Luzon

REGION 3

REGION 5

Visayas

REGION 4

REGION 8

Palawan

REGION 6

REGION 10

REGION 7

REGION 9

REGION 12

REGION 11

Mindanao

〰 Superregion	less than 20% forest cover
〜 Region	between 20% and 40% forest cover
⌒ Province	greater than 40% forest cover

50 km

N

Fig. 1. Forest cover of the Philippines in 1980, showing provinces and regions

and year of publication. An additional source of information on vegetative cover is the World Bank–funded SPOT survey undertaken by the Swedish Space Corporation (SSC). Both sources of forestry data will be discussed in detail below.

Deforestation has attracted a great deal of professional and public attention in the Philippines. The major environmental effects appear to be the following:

1. Soil erosion on areas that have been deforested. The major on-site effect has been land degradation and the major off-site effect has been sedimentation of rivers. As a result of sedimentation, major hydroelectric dams will experience a shortening of their useful life span, deposition in irrigation systems has increased, floods have worsened, dry-season flows have lessened, and siltation of coral reefs has occurred (Briones 1986; Finney and Western 1986; Hodgson 1988; Pantastico and Metra 1982; Porter and Ganapin 1988; Revilla 1988; Segura et al. 1977). The World Bank (1989a) claims that soil erosion is the most serious environmental problem facing the Philippines today.

2. Destruction of wildlife (Agaloos 1984; Grimwood 1975; Johnson and Alcorn 1989; Talbot and Talbot 1964; World Bark 1989a).

3. Loss of secondary forest products such as gums, resins, and rattan (Agaloos 1984; DeBeer and McDermott 1989; Talbot and Talbot 1964).

In addition to these environmental effects, three other considerations are worthy of mention. First, commercial logging and the spread of agricul-

Key to fig. 1:

1	Ilocos Norte	20	Quezon	38	Samar	56	Lanao del Norte
2	Kalinga-Apayao	21	Bataan	39	Eastern Samar	57	Lanao del Sur
3	Cagayan	22	National Capital	40	Leyte	58	Misamis Oriental
4	Abra		Region (Manila)	41	Batanes	59	Camiguin
5	Ilocos Sur	23	Rizal	42	Southern Leyte	60	Surigao del Norte
6	Mt. Province	24	Laguna	43	Bohol	61	Agusan del Norte
7	Isabella	25	Cavite	44	Cebu	62	Surigao del Sur
8	La Union	26	Batangas	45	Negros Oriental	63	Bukidnon
9	Benguet	27	Camarines Norte	46	Negros Occidental	64	Agusan del Sur
10	Ifugao	28	Occidental Mindoro	47	Iloilo	65	Davao del Norte
11	Nueva Vizcaya	29	Oriental Mindoro	48	Capiz	66	Davao Oriental
12	Quirino	30	Marinduque	49	Aklan	67	Maguindanao
13	Pangasinan	31	Camarines Sur	50	Antique	68	North Cotabato
14	Nueva Ecija	32	Catanduanes	51	Palawan	69	Sultan Kudarat
15	Aurora	33	Albay	52	Siquijor	70	South Cotabato
16	Zambales	34	Sorsogon	53	Zamboanga del	71	Davao del Sur
17	Tarlac	35	Romblon		Norte	72	Basilan
18	Pampanga	36	Masbate	54	Zamboanga del Sur	73	Sulu
19	Bulacan	37	Northern Samar	55	Misamis Occidental	74	Tawi-Tawi

ture have led not only to deforestation but to the destruction of the home-lands of numerous ethnic groups in the Philippines who live in the forests (Anti-Slavery Society 1983; Episcopal Commission of Tribal Filipinos 1982; Headland 1987; *Southeast Asia Chronicle* 1979). Second, since deforestation has been so extensive in the past forty years, there is the possibility that the Philippines may face a wood shortage by the year 2000. If so, one of the effects of deforestation would be the disruption of the forestry-based industries and dislocation of workers (PREPF 1980). Third, deforestation has been very prof-itable for a small group of people in the Philippines (Diaz 1982). I will argue that the profits from deforestation have reinforced the power of the better-off members of society and led to large-scale corruption in the forestry sector. The present research agrees with the overall conclusion of Porter and Ganapin (1988) that the process of deforestation in the postwar Philippines has helped produce a degraded environment and impoverished socioeconomic system.

A detailed history of forest cover in the Philippines from pre-Spanish times to the present day would be difficult to recreate. Not only were forest records incomplete throughout almost the entire period but many of the records that did exist have been destroyed. Almost all of the Spanish forest records were lost in a fire in Manila in 1897 (Tamesis 1948) and the records of the BF in Manila and the College of Forestry in Los Baños were destroyed dur-ing fighting in 1945 (Sulit 1947).

Unfortunately, lost or destroyed forestry data are not restricted to the period before 1945. As an example, from 1954 to 1961 forest survey crews, working primarily in Mindanao, inventoried approximately 1.65 million ha of forestland (Agaloos 1976; Serevo et al. 1962). Information was collected on forest area, timber volume, and species composition and would provide a valuable benchmark to assess changes in the forest resource since the inven-tories were completed. However, according to staff of the FMB in Manila, these surveys cannot be found. In addition, librarians at the FMB and De-partment of the Environment and Natural Resources (DENR) have no knowledge of these reports. In short, seven years of work by skilled survey crews seems to have disappeared.

Another difficulty concerns the accuracy of data from the FMB, the offi-cial source of all forest-cover data in the Philippines. As will be discussed be-low, there is reason for serious doubt as to the reliability of much of the offi-cial government data on forestry, and the discrepancies between FMB data and more objective data are often quite large. An additional consideration, to be discussed below, is that there appear to be cases where forestry statistics have been deliberately manipulated. In short, any attempt to describe forest cover in the Philippines in the twentieth century must confront the fact that (1) many records are incomplete, (2) many data that have been collected have been lost, (3) the data are quite often of dubious quality, and (4) there is sub-stantial circumstantial evidence to indicate deliberate manipulation or destruction of data by government officials. I have personally encountered all of these problems.

Philippine forests are usually divided into six forest types: dipterocarp, molave, beach, pine, mangrove, and mossy. The molave forest is a dry, monsoonal forest which is found only in parts of the western Philippines (central Luzon, Mindoro, and Palawan). According to Agaloos (1984), the molave forest type makes up only 3% of the total forest area of the Philippines. It is usually included in the category dipterocarp (Umali 1981). The beach forest used to occur on coastal areas and was a transition between the mangrove forest and other forest types inland. For all practical purposes, beach forests no longer exist in the Philippines (Agaloos 1984) or Southeast Asia (Whitmore 1984). There are two types of pine native to the Philippines: Benguet pine (*Pinus kesiya*), found in northern Luzon, and Mindoro pine (*Pinus merkusii*), found in parts of Mindoro and western Luzon. Altogether pine forests occupy 2,390 km^2 (P-GFI 1988).

Mangroves are restricted to coastal fringes and tidal flats and now occupy approximately 1,391 km^2 (P-GFI 1988). They have been subjected to intense pressure because their woods are valuable for fuel (charcoal) and cutch, and many of them have been converted to fishponds (Gillis 1988; Johnson and Alcorn 1989). In fact, according to the SSC (1988), fishponds derived from mangroves now cover 1,952 km^2. Alcala (1987) claims that mangrove forests occupied 5,000 km^2 at the turn of the century.

The mossy forest (also referred to as mountain or cloud forest in the literature and unproductive forest by the FMB) is found at higher elevations, usually above 1,800 m. It is a stunted forest and has no commercial value (Agaloos 1984; Weidelt and Banaag 1982). It is distributed throughout the Philippines and its primary role is in water- and soil-holding functions. According to the P-GFI (1988) it presently covers 11,374 km^2.

The major forest type in the Philippines is composed mainly of species belonging to the family Dipterocarpaceae. They account for more than 90% of all commercial forest products in terms of economic value (Agaloos 1984). The history of deforestation in the Philippines is primarily the history of the decline of the dipterocarp forest. While destruction of mangroves has been rapid and dramatic, the area involved is insignificant compared to the area of dipterocarps that has been destroyed. This does not deny the ecological role that mangroves play in protecting coastlines and as an important component of the food chain. Unless otherwise stated, forest area in this study will refer to the total forest cover of all forest types.

The task of evaluating what is happening to forest cover is complicated by the issue of land classification. An important distinction is made in the Philippines between officially designated forestlands and lands which are alienable and disposable (A&D). A&D lands are former forestlands which have been declared nonforestlands by the government (FMB) and are eligible

Table 3.
Land classification in the Philippines, 1958 and 1987 (km^2)

	1958	1987
A&D	116,296	141,524
Forestland		
Classified	34,595	150,116
Unclassified	146,522	8,813
Total	297,413*	300,000*

Sources: NEC (1959, table 2); *PFS* (1987, table 1.1).

*In the postwar period, the official total land area of the Philippines has varied between 297,000 and 300,000 km^2. The 300,000 km^2 figure is now generally accepted and has been used in this study. The discrepancy most likely arises from including the area of inland waterways.

for private ownership. The major criteria for distinguishing between the two categories is slope: lands with a slope of 18% or greater are generally considered to be forestlands (Revilla 1981). Almost all forests are now on forestlands. According to the P-GFI (1988), of the 6.5 million ha of forest cover in 1988, only 108,700 ha (1.7% of all forests) were on A&D land.

A widely discussed issue in postwar Philippine forestry has been the rate at which unclassified forestlands have been declared to be either forest or A&D land. Most observers have argued that the process has been slow and has furthered deforestation because unclassified forestlands have been accessible to anyone who wanted to use them (Basa 1982; Segura et al. 1977; Serevo et al. 1962; Utleg 1958). Table 3 presents land classification data for the years 1958 and 1987.

It is obvious from table 3 that land classification in the Philippines has primarily meant changing unclassified forestland into classified forestland; however, most classified forestland is without forest cover. According to the P-GFI (1988), of the 15.9 million ha of official forestland in 1987, only about 40% was covered with forests. This means that the FMB is technically in charge of 55% of the total land area of the Philippines, although forests cover less than 25% of the total area. Since the sizes of the forest and A&D categories have changed continuously since 1950 as forestlands have been reclassified as A&D, it is impossible to trace changes in forest cover over time for the two land classifications. Forest cover on forestlands for two separate dates can be compared, but the areal extent of the forestland category would most likely have changed in the intervening time period. In effect, forests on forestland could have been lost through actual deforestation or through reclassification into A&D land.

An additional consideration regarding forest cover has to do with logging concessions and logging production as reported in the annual issues of *PFS* (1968 to present). Keeping in mind that the total area of the Philippines is 300,000 km^2, concession area increased from 45,000 km^2 in 1960 to an average of 98,000 km^2 during the 1971–77 period. Since then, concession area has steadily declined to 56,000 km^2 in 1987, although this still represents 19% of total land area. Log production increased from 6,315,000 m^3 in 1960 to an average of 10,500,000 m^3 during the 1968–75 period. Since then it has steadily decreased to approximately 4,000,000 m^3 in 1987. The percentage of log production that was exported increased from approximately 50% in 1960–61 to an average of 78% during the 1967–72 period. In 1987, 5% of log production was exported.

As will be discussed below, I do not put much faith in official forestry statistics from the Philippines. The log production and export figures presented above must be considered to be minimum figures. They do document, however, the rapid increase in officially sanctioned logging which occurred in the late 1960s and early 1970s and the high degree of dependence on foreign markets (primarily Japan).

Forest Cover and Exploitation, 1876–1950

Deforestation in the Philippines is not restricted to the twentieth century. Wernstedt and Spencer (1967) report that forest cover declined from approximately 90% of total land area at the time of the first contact with the Spanish in 1521 to approximately 70% by 1900. The major causes were most likely the steady increase in population during this period and the spread of commercial crops (primarily abaca, tobacco, and sugarcane) as the Philippines was slowly integrated into the world economic system (Lopez-Gonzaga 1987; McLennan 1973; Roth 1983; Westoby 1989). The spread of commercial crops was so extensive that by the second half of the nineteenth century, the Philippines began to import rice (Krinks 1974; Reynolds 1985).

Reliable statistics on forest cover before 1950 do not exist; thus, any discussion of forest cover and its decline during this period must be based on estimates made by contemporary observers. In addition, comparisons between the various estimates are difficult to make because (a) definitions of forest cover and (b) methods of evaluation are rarely made clear. The summary presented in table 4 is thus meant to be broadly indicative of total forest cover at the national level.

The figures presented in table 4 indicate a decline in forest cover from 70% in 1900 to just under 60% in the late 1930s. World War II, particularly the occupation of the Philippines by the Japanese from 1942 to 1945, severely disrupted Philippine society. Two main influences affected forest cover: on the one hand, logging production declined, and on the other, large numbers of Filipinos took refuge in the forests, where they practiced farming (Sulit, 1947). The overall effect of these two factors on forest cover is unknown.

Logging increased rapidly after 1945 and was back to prewar levels of production by 1949 (Poblacion 1959; Tamesis 1948). In addition, farming in forests increased after the war as a result of the continuing shortage of food (Sulit 1963; Tamesis 1948). The overall extent of deforestation in the immedi-

Table 4.
Percentage forest cover, 1876–1950

Date	Forest cover (%)	Source
1876	68	U.S. Bureau of the Census (1905)
1890	65	BF (1902)
1900	70	Wernstedt and Spencer (1967)
1903	70	U.S. Bureau of the Census (1905)
1908–10	50+	Whitford (1911)
1910	66	Zon (1910)
1911	64	Talbot and Talbot (1964)
1918	68	Census Office of the Philippine Islands (1920)
1919	67	Wernstedt and Spencer (1967)
1923	50	Zon and Sparhawk (1923)
1929	57+	Borja (1929) (commercial forests only)
1934	58	Revilla (1988)
1937	57	Tamesis (1937)
1937	58	Pelzer (1941)
1939	60	FAO (1946)
1943	60	Dacanay (1943)
1944	60	Allied Geographical Section (1944)
1945	66	Hainsworth and Moyer (1945)
1948	59	FAO (1948)
1948	59	Tamesis (1948)
1950	55	Myers (1984)

ate postwar period is not known. Tamesis (1948), then director of the BF, estimated that commercial and noncommercial forests comprised approximately 59% of total land area in 1948. However, since Tamesis had reported 57% coverage in 1937, it is difficult to see how forest cover could have increased by 2% from 1937 to 1948. Myers (1984) suggested a figure of 55% for 1950, but a figure closer to 50% is probably more accurate (to be discussed below).

Forest Cover and Exploitation, 1950–88

In a study for the U.S. Mutual Security Agency which relied on data from the BF, Gooch (1953) stated that 55.4% of the country was under forest cover in 1951 (including mangroves and marshes). Amos (1954), who was then director of the BF, uses the same figure to describe forest cover in 1953. The same figure is also given in Tamesis (1956). In short, between 1951 and 1956, according to the sources above, no deforestation occurred in the Philippines. The FAO (1955) presents data for 1953 which indicate that the Philippines was 54.9% forested, but this data came from the BF and cannot be considered an independent source.

Keith (1956), in an FAO report to the Philippine government, noted that no inventory of Philippine forests had yet been conducted and recommended that one be undertaken immediately. Bedard (1958) echoed Keith's observation by noting that the latest data from the BF were from 1954 and quoted the same 55.4% figure given by Gooch (1953) above. He also noted that the BF had initiated a nationwide forest inventory based on controlled sampling on the ground. Bedard claimed that 2 million ha had been inventoried already, and I assume that this is the 1954–61 survey in Mindanao referred to in the introduction to this chapter. As I mentioned, all of the results of this inventory appear to have been lost.

Lansigan (1959) seems to be the first Filipino forester to present evidence documenting a serious timber drain. His source was an unpublished report of the Inter-Agency Committee on Up-Dating Forestry Statistics (1959). Unfortunately, copies of this report could not be found at the FMB, DENR, the College of Forestry in Los Baños, or any government agency or library. However, a perusal of the Lansigan article indicated that the the Inter-Agency Committee used as its primary source a report from the National Economic Council (NEC) (1959).

The NEC report was an assessment of forest and agricultural lands. It was not derived from remotely sensed data; rather, it was based on reports submitted by district foresters and the BF central office. Based on the data they gathered, the NEC expressed alarm at the rapid rate of deforestation occurring in the Philippines. The difference between the 1953 data on forest cover by the BF and the 1957 data used by the NEC is presented in table 5.

The NEC data indicate a decrease in the area of commercial forests by approximately 2.1 million ha and in noncommercial forests by 0.6 million ha. They also introduce the new category of brushland. Interestingly, the decrease in commercial forests between 1953 and 1957 (2,085,720 ha) is just about equal to the area of brushland reported for 1957 (2,077,230 ha).

If the NEC data are correct, they would indicate a rapid conversion of commercial and noncommercial forests to a degraded or deforested state (brushland or open land), highlighting the importance that a category such as brushland plays in forest-cover statistics. That the NEC evaluation was considered official by the Philippine government is indicated by the fact that the FAO (1960) reported Philippine forest cover as 13,173,000 ha, almost exactly the total of commercial and noncommercial forests presented in table 4 by the NEC for 1957 (13,171,400 ha). This is equivalent to 44.3% of total land area.

In 1959, Tom Gill, an American forester, conducted a two-month study of Philippine forests for the NEC and the International Cooperation Administration. In his initial report (1960a) and later writing (1960b), Gill deplored what he considered to be the rapid destruction of Philippine forests. His study received a great deal of attention in the Philippines (Lansigan 1959; Society of Filipino Foresters 1959–60) and, although he did not present any statistical data on forest cover, it is clear from his comments that his views coincide more with the NEC than the BF.

Table 5.
Philippine forest cover data for 1953 and 1957

Vegetative cover*	1953 (BF) (in ha)	% Total land area	1957 (NEC) (in ha)	% Total land area
Commercial forest	11,400,000	38	9,300,000	31
Noncommercial forest	4,500,000	15	3,800,000	13
Brushland	—	—	2,100,000	7
Swamps and marshes	600,000	2	700,000	2
Total	16,500,000	55	16,000,000	54

Sources: Adapted from Amos (1954), NEC (1959).

*Commercial forests are forests of commercial species in which the volume of trees greater than 30 cm at diameter breast height is greater than 40 m^3/ha. Brushland is defined as areas dominated by shrub and brush.

The concerns of the NEC (1959) and Gill (1960a, 1960b) were repeated by another American researcher, Shirley, in 1960. Soon after, Serevo et al. (1962) and Tamesis (1963) presented data indicating that forest cover (including swamps) was approximately 47% of total land area in the early 1960s. At about the same time, the FAO (1963) published data indicating that forest cover was around 40%, and since this figure included lands officially designated as forestlands but without forest cover, the actual forest cover was presumably less than 40%.

The first national forest inventory was undertaken from 1962 to 1968. The Philippine Air Force took strip aerial photography of the entire Philippines every 15 km at a scale of 1:15,000 (Aranas 1973). The output of this project was five publications which analyzed forest cover for Palawan and all of Mindanao (Agaloos 1964a, 1964b, 1965a, 1965b; Agaloos and Santos 1968). Inventory results were compared to black and white aerial photographs taken by the U.S. Army at a scale of 1:46,000 from 1946 to 1953 (Roque 1978) and changes in forest cover for the entire country were calculated. Overall rates of deforestation were presented for Palawan and Mindanao in the reports cited above and for Luzon and the Visayas in various issues of PFS.

Unfortunately, the individual studies for Luzon and the Visayas were never published, and thus the extent of forest cover in those areas was never made public. Nilsson et al. (1978) state that a report covering the island of Mindoro was available, but neither the FMB nor I was able to find it. Singh (1979) notes that the ground tally sheets for this inventory were available at the time he was writing; if true, this means that the inventory results were lost sometime between 1979 and 1988. Thus, a national forest inventory which took six years to complete did not produce any data on national forest cover because the work for Luzon and the Visayas was never published and appears to be lost.

The annual rates of deforestation (in hectares) determined by the first national forest inventory were: Mindanao (–91,564), Palawan (–4,191), Visayas

(–31,839), Luzon (–76,311), and nationwide (–203,905). The Mindanao data cover the years 1952–63 and the Palawan data 1948–64. The inventory for the Visayas ended in 1968 and for Luzon in 1967, but information on what year the 1946–53 aerial photographs were taken in those regions is not available. These deforestation rates, with some modification, will be used below to project forest cover back to 1950 after forest cover for 1969 has been determined.

The importance of the deforestation rates presented above cannot be underestimated. They provided the basis for almost all official forest-cover statistics after 1971. In 1984, official land-use statistics were based on projections using these deforestation rates (*PFS* 1984), and as late as 1987, land-use data for three of the twelve regions in the Philippines were based on projections using these rates of change (*PFS* 1987). As the discussion below will indicate, these rates of change continued to be used in spite of strong evidence that they were inaccurate.

The Presidential Economic Staff (1968) issued a report based on the completed inventories for Palawan, Mindanao, and Luzon and on estimates for the nearly completed survey of the Visayas. Their findings indicate that forest cover in 1967 was 16,663,090 ha, or 55.5% of total land area. This figure is obviously very difficult to reconcile with any of the data discussed in this chapter so far. It is also the figure reported in the 1968 *PFS* in its first year of publication. To put the above figures in perspective, Talbot and Talbot (1964) calculated that forest cover in the Philippines in the early 1960s was approximately 35% of total land area. The difference between the two estimates is approximately 6 million ha, or 20% of the total land area of the Philippines.

The second national forest inventory was taken between 1965 and 1972. Aerial photography was taken of 89% of the land mass of the Philippines by Certeza, a private company (Agaloos 1976). The results were published as tables 6 and 7 in the 1973 *PFS*. Most researchers take 1969 as the midpoint of the study, but even this is open to doubt (Revilla 1988). In addition, the data presented in tables 6 and 7 often do not agree with the data presented on table 2 of the same report. Given these caveats, a comparison of the 1972 and 1973 *PFS* is revealing.

The 1972 *PFS* indicated 15,671,104 ha of forest, while the 1973 *PFS* reported 13,893,963 ha. In other words, between 1972 and 1973, according to the official source on Philippine forestry statistics, almost 1.8 million ha of forest cover were lost. This most likely explains the unusual history of the 1973 *PFS*. According to sources who prefer to remain anonymous, the 1973 *PFS* was withdrawn almost as soon as it was released; the declaration of martial law in September 1972 may have made this easier. The FMB and DENR do not have a copy of this report and it was only by serendipity that I was able to locate a copy in one of the libraries of the Department of Agriculture.

According to Aranas (1973), Certeza, the firm which took the aerial photographs for the 1965–72 forest inventory, published a progress report in

1972 based on these photos. Unfortunately, neither the FMB, DENR, the College of Forestry in Los Baños, nor any government library has this report and, in fact, Certeza does not have a copy (Certeza and Co. 1988).

In actuality, the 1.8 million hectare difference between the 1972 and 1973 *PFS* understates the extent of the problem. Since we have taken 1969 as the midpoint of the inventory data presented in the 1973 *PFS*, the correct comparison should be with the 1968 *PFS*. If this comparison is made, the difference increases to 2.8 million ha, which is equivalent to 9.2% of the total land area of the Philippines. The second national forest inventory seems to be a good example of what Callaham and Buckman (1981, 56) have observed: "In some cases the efforts to provide an inventory of forest resources have been thwarted because accurate information might prove embarrassing or cause undesirable changes in the viewpoints of policy makers."

The quality of the data from the second national forest inventory is difficult to judge. The entire output of this seven-year study takes up two tables and five pages of the 1973 *PFS*. In addition, while the aerial photos cover 15.1 million ha, the area not covered (presumably because of cloud cover) equals 2.3 million ha, or roughly 15% of the total area. There is also some doubt as to how accurately the photo interpretation was conducted. Since the P-GFI (1982–88) did not have the results of this national forest inventory available to them (they could not find a copy of the 1973 *PFS*), they were forced to calculate forest area in 1969 by remeasuring all of the 1969 forest resource condition maps (FRCMs), which were the major output of the second national forest inventory. Their work uncovered numerous classification errors in the 1969 FRCMs (see various regional inventories of the P-GFI project, 1982–88; Coppin and Lennertz 1986). In addition, the second national forest inventory covered only official forestlands; forests on A&D lands were not included.

The statistics reported in the 1973 *PFS* have been used by Filipino foresters to calculate rates of forest decline in the Philippines (Ganapin 1987; Revilla 1983, 1988). Revilla (1983) has interpreted the 1973 *PFS* to mean that in 1969 the Philippines was approximately 33.5% forested (not including reproduction brush), but this does not include forest cover on private land. Given the uncertainties of these data, I decided to accept the findings of the P-GFI for 1969.

The P-GFI shows 104,565 km^2 of forest cover in 1969, and the first national forest inventory determined a deforestation rate of 2,039 km^2 per year from approximately 1950 to the mid-1960s. Deforestation increased in the 1960s and 1970s (to be discussed below), so it is reasonable to assume that the 2,039 km^2 per year is an underestimate of deforestation up to 1969, since the regional inventories were completed between 1962 and 1968. If we assume that deforestation was on average 10% higher during the 1950–69 period (a rate of 2,243 km^2 per year), then forest cover in 1950 would have been 147,180 km^2. This is equal to 49.06% of total land area and is a reasonable estimate of national forest cover for that year.

If forest cover in 1950 was approximately 50%, this necessarily means that deforestation was occurring throughout the late 1930s and 1940s. It also means that government estimates regarding forest cover in the 1940s were incorrect. If the deforestation rate of 2,243 km^2 is projected from 1969 back to 1957, it yields a figure of 131,480 km^2. This is 44.2% of total land area (using the NEC figure of 297,412 km^2 for total land area) and agrees quite well with the 44.3% figure suggested by the NEC (1959). While the results of the NEC study were not based on aerial photos, their apparent accuracy could mean that it was an honest evaluation of forest resources at the time.

The first satellite-assisted forest inventory was conducted jointly by the BFD and the General Electric Co. using LANDSAT images primarily from 1972 and 1973, with 1973 taken as the midpoint (General Electric Co. and Department of Natural Resources 1977; Lachowski et al. 1979). Results from this inventory showed that forest cover was 114,027 km^2, or 38.0% of total land area in 1973. The authors admit that their calculations provide only a rough estimate of forest cover. Difficulties encountered during this inventory include: cloud cover, overlap of images, shadows in areas of high relief, different dates of images, and problems with interpreting some of the classification categories. As a result of these drawbacks, the "forest, partial closure" category, which equals 38,176 km^2, may include large areas of degraded forests, and the "forest obscured by clouds" category, which equals 15,199 km^2, must be interpreted with care. The reliability of this inventory will be discussed in more detail in chapter 8.

An interesting aspect of this inventory (which was actually the third national forest inventory conducted in the Philippines) is that, although the authors mention the first inventory (1962–68), they never once mention the second inventory (1965–72). This tendency to either ignore or not be aware of previous work is commonplace in the Philippines and is shared by Filipino and foreign researchers. The recently completed Philippine-German forest inventory is actually the fourth national forest inventory conducted by the Philippine government and is based on work initiated by FAO consultants in the late 1970s (Nilsson et al. 1978; Palo 1980; Singh 1979).

Between 1976 and 1980, five estimates of national forest cover in the Philippines based on LANDSAT imagery were published:

1. Eckholm (1976) stated that satellite imagery showed forests to be less than 20% in 1976. However, since no reference is given, it is impossible to verify this claim.

2. Bonita and Revilla (1977), on the basis of a visual interpretation of a composite LANDSAT image, calculated that forest cover in 1976 was 30% of total land area.

3. Bruce (1977), on the basis of 1973–75 LANDSAT imagery, claimed that forest cover was 89,303 km^2, or approximately 30% of total land area in 1974.

4. Scott (1979), on the basis of 1976–78 LANDSAT imagery, claimed that forest cover was approximately 25% of total land area in 1977.

5. The Forestry Development Center (FDC) (1985), on the basis of a visual interpretation of LANDSAT imagery, estimated that adequately stocked forests were 79,620 km^2, or 25.9% of total land area in 1980.

The official country report of the Philippines to the World Forestry Congress in 1978 claimed that forests covered 170,300 km^2, or 56.76% of the total land area of the Philippines (Myers 1980). As remarkable as this claim appears in light of the information presented above, an equally incredible claim was made the next year. In 1979, the Natural Resource Management Center (NRMC) (1980) of the Ministry of Natural Resources estimated that forest cover was 165,000 km^2, or 55% of total land area. In light of the studies presented above, this finding by a government agency is quite surprising and raises serious questions as to the honesty and purpose of the exercise. It is interesting that the experts involved in this exercise are not named. On the other hand, Nilsson et al. (1978) pointed out that estimates of forest cover were uncertain and suggested a range of 80,000 to 130,000 km^2, or 27 to 43% of total area. Partly as a result of this wide range of estimates regarding forest cover, Palo (1980) was sent by the FAO to design a plan for Philippine forestry statistics.

Since 1980, two major national forest/vegetation inventories have been conducted: the P-GFI (1982–88) and a World Bank–funded study by the SSC (1988). In effect, these are the fourth and fifth national forest/vegetation inventories conducted or participated in by the Philippine government since the early 1960s.

The SSC study relied exclusively on SPOT satellite data, and approximately 98% of total land area was classified. A total of 187 separate images were used, with the first image received in March 1987 and the last in February 1988. Most of the images are from 1987 and I accept 1987 as the year of completion. Ground truthing (the acquisition of reference data on the vegetation of the Philippines) was conducted over a six-week period in the 1987 dry season and consisted of aerial reconnaissance and ground surveys. Ground surveys were conducted only in northern and southern Luzon and Cebu. The SPOT images were interpreted visually.

In the SSC study, land cover was divided into four groups: forest (five classes), extensive land use (three classes), intensive land use (seven classes), and nonvegetated lands and other areas including marine areas (nine classes). "Forest is defined as forest trees and reproductive brush areas with less than 10% of cultivated and other open areas" (SSC 1988, 17). The forest classes are closed dipterocarp (canopy closure of mature trees > 50%), open dipterocarp (canopy closure of mature trees < 50%), mossy, pine, and mangrove.

Their results indicate that forestlands occupy 25% of the total area classified. This is correct but misleading, because if nonvegetated and unclassified land areas are included, then forest as a percentage of total land area decreases to approximately 23.7%. The difference of 1.3% represents 3,900 km².

The P-GFI is the continuation of an FAO-assisted inventory which covered regions 10 and 11 in Mindanao from 1979 to 1982. Starting in 1983 and ending in 1988, the remaining ten regions were inventoried by the BFD/FMB with the assistance of Deutsche Gesellschaft für Technische Zusammenarbeit, the German overseas development agency. Approximately 80% of all forestlands were mapped using aerial photography from the 1980s and 20% were mapped using LANDSAT or SPOT imagery. After mapping, dipterocarp and pine forests were sampled in the field to determine stand structure, species composition, and timber volume. In addition, rattan, bamboo, and erect palms were inventoried. A total of 2,627 sample clusters were taken nationwide. The P-GFI identified nine forest classes: old-growth dipterocarp (no signs of commercial logging), residual dipterocarp (cut over dipterocarp forest), submarginal (tropical forest composed of noncommercial species), mossy, closed pine forest (crown cover > 30%), open pine forest (crown cover between 10 and 30%), mangrove (old-growth and reproduction), and forest plantation (Coppin 1984).

Rates of deforestation for each province were determined by comparing forest cover in the 1980s with the 1969 FRCMs or, in some cases, the results of the first national forest inventory. These rates of deforestation were then projected for each region to arrive at a national figure for forest cover in 1988. The results, projected back to 1987, indicate total forest cover of 66,700 km² in 1987 (22.2%) as compared to 71,046 km² (23.7%) for the SSC. While the results of the two inventories are fairly similar at the national level, a closer examination reveals some rather large differences.

Table 6 presents inventory results for the dipterocarp, pine, mossy, and mangrove forests at the national level and also the results for selected provinces and regions. I assume that the one-year difference between 1987 and 1988 is not significant.

The discrepancy between the two sets of figures could be caused by several factors. First, the stratification of the category of dipterocarp is different: percentage canopy cover for the SSC, and logged/nonlogged for the P-GFI. Second, forests that are classified as submarginal by the P-GFI and not included in the dipterocarp category may be included in the SSC category of open dipterocarp. Third, classification errors may have occurred in one or both surveys. In particular, it is now agreed that the SSC (1988) seriously underestimated the extent of mossy forest (World Bank, 1989a). Regardless of the cause of the difference, it is obvious, for instance, that old-growth dipterocarp

Table 6.
SSC (1987) and P-GFI (1988) estimates of forest cover (km^2)

	SSC (1987)	P-GFI (1988)	
Dipterocarp, closed	24,345	9,883	Dipterocarp, old-growth
Dipterocarp, open	41,940	34,128	Dipterocarp, residual
Total	66,285	44,011	Total
Pine	812	2,388	
Mangrove	1,494	1,391	
Mossy	2,455	11,374	
Mangrove, Sulu	—	147	
All forests, Bohol	251	147	

Sources: SSC (1988); P-GFI (1988).

in the P-GFI cannot be equated with closed dipterocarp in the SSC inventory. In short, each inventory presents a very different picture of the extent and composition of forest types at the national level. The differences are so great as to raise very serious questions as to which inventory is the more appropriate. At the provincial and regional level, the differences can be equally striking. In short, by forest class and geographical area, the results of the two inventories are not equivalent.

The goal of the SSC inventory was a mapping of the natural conditions (vegetation) of the Philippines, while the goal of the P-GFI was a detailed forest inventory. The advantages of the SSC data are threefold: they are recent (1987), they were gathered during a short time period (less than one year), and they are homogeneous. The major difficulty arises from the fact that the SSC inventory was the first study using high-resolution satellite imagery (SPOT) on such a large scale (an entire tropical country), and the ground truthing was inadequate. As mentioned above, ground surveys were conducted only on Luzon and Cebu and, since the purpose of the survey was to inventory vegetative cover, they were concerned not only with forest cover but also with intensive and extensive agriculture. A total of seven ground surveys were conducted, and it would appear that the average time spent on each was between two and three days (SSC 1988, app. 3). A total of nineteen air reconnaissances were conducted and each took one day (or less). As an example, the islands of Palawan (14,896 km^2) and Mindanao (101,999 km^2) were covered by air reconnaissance which took one and three days respectively. In short, given the extensive nature of the project, the ground truthing, in comparison, seems minuscule.

Two other points are worth mentioning: first, while each of the seven ground surveys was accompanied by at least one Filipino expert from the NRMC, only six of the nineteen air surveys were accompanied by an NRMC staff member; second, all air and ground surveys were done in the dry season, but it may be the case that some of the images are from the wet season. A logical question is whether or not ground truthing in the dry season can provide enough information to interpret images taken during the wet season. As an example, according to a P-GFI staff member (personal communication 1988), on a wet-season image it is impossible to separate molave from dipterocarp forest. Add to this example the difficulties the SSC encountered in differentiating between mossy and closed-canopy forest (mossy forest has been underestimated and closed-canopy dipterocarp forest overestimated; World Bank, 1989a), and the relative lack of ground truthing could present major problems in interpretation of the spectral signatures of the SPOT images.

By contrast, the P-GFI took place over a period of nine years (including the FAO period from 1979 to 1982) and included over 2,600 detailed field samples. In addition, 80% of all forests were inventoried with aerial photographs with a greater resolution than the SPOT imagery. At the same time, a major shortcoming of the P-GFI inventory is that the regional inventory data from the 1980s has been projected to arrive at a figure for 1988 using rates of deforestation established for the period from 1969 to the 1980s. The year of completion of the inventory and the corresponding regions are as follows: 1980 (regions 10 and 11), 1981 (regions 1, 2, and 3), 1984 (region 4, not including Palawan, and regions 5, 8, and 12), 1985 (Palawan), 1987 (regions 6, 7, and 9). As a result, for example, the data on forest area for regions 10 and 11 were already eight years out of date when the nationwide inventory was published in 1988. For the twelve regions plus Palawan, the average year of completion was mid-1983. Thus, on average, regional forestry data have been projected four and a half years forward to 1988. Needless to say, the assumption that rates of deforestation are the same in the 1980s as they were in the 1970s is a major one and may be incorrect.

Although the two inventories are fairly close in terms of national forest cover in 1987, the discrepancy is substantial enough to have a large effect on the calculation of rates of deforestation. Since the difference between the two inventories for the year 1987 is equal to 4,346 km^2, they will produce very different rates of deforestation for the 1980–87 period.

One of the major differences between the SSC study and the P-GFI is that the SSC study and the World Bank (1989a) report based on it make very little effort to compare their results regarding forest cover with any other earlier work on forest cover. The World Bank (1989a, 113) claims that this was not possible "due to the weaknesses of the Philippine statistical system." The P-GFI, on the other hand, compares forest cover in the 1980s with forest cover indicated on the 1969 FRCMs.

While I am sympathetic to the observation of the World Bank that Philippine statistics leave a lot to be desired, and while I recognize the difficulties in comparing forest surveys based on different definitions and methodologies, I am certain that this is a necessary first step to truly understanding the processes of land-use change in the Philippines. The World Bank (1989a, 113) is aware of this and notes that "a special effort may be necessary to make present analytical work comparable with earlier work." Unfortunately, they have not undertaken such an effort themselves. Since the SSC/World Bank study was the first to use SPOT imagery for an entire tropical country, such an endeavor on the the part of the authors could have helped a great deal to clarify the strengths and weaknesses of various forest inventories.

In short, my study was confronted with a completely unexpected result: even forest inventories which are fairly close in their estimates of national forest cover can lead to very different views of forest cover at the subnational level. Since the P-GFI is based on imagery with a higher resolution than the SPOT data, is the result of extensive ground truthing by a team of experts much more familiar with Philippine conditions than the SSC, and provides forest-cover data for two separate years, I decided to use the P-GFI results for the statistical analysis in chapter 7. (See fig. 1 for the distribution of forest cover at the provincial level for 1980 in the Philippines.) The larger question of what role satellite remote sensing can play in forest inventories in the tropics will be taken up in chapter 8. Table 7 summarizes the results of the inventories discussed so far.

Table 7.
National forest cover, 1950–87 (percentage of land area)

Date	Percentage	Source
1950	49.1	Projection from 1969[a]
1957	44.3	NEC (1959)[ac]
1969	33.5	*PFS* (1973) as interpreted by Revilla (1983)[b]
1969	34.9	P-GFI (1988)[a]
1973	38.0	Lachowski et al. (1979)[ad]
1974	29.8	Bruce (1977)[a]
1976	30.0	Bonita and Revilla (1977)[ae]
1976	< 20	Eckholm (1976)[a]
1977	25.0	Scott (1979)[a]
1980	25.9	FDC (1985)[a]
1980	27.1	P-GFI (1988)[af]
1987	23.7	SSC (1988)[ag]
1987	22.2	P-GFI (1988)[ah]

[a]Includes forest and A&D lands.

[b]Forestlands only.

[c]Does not include brushlands or marsh/swamps.

[d]Does not include nonforest wetlands. Brushland may be included in the categories "forest, partial closure" and "forest obscured by clouds." Collectively, these two categories comprise 53,375 km^2, or approximately 48% of total forest area.

[e]Since the original figures included approximately 10% brushland (Revilla 1988), the total was reduced by 10%.

[f]Calculated by projecting P-GFI rates of deforestation back to 1980.

[g]Does not include land area not classified.

[h]The P-GFI figures for 1988 were projected back to 1987.

Table 8.
Superregional forest cover, 1950–87 (km^2)

Year	Mindanao	Palawan	Visayas	Luzon	Source
1950	61,576	10,703	18,066	56,838	Projection
1957	52,067	11,313	15,436	52,899	NEC
1969	42,439	9,827	11,412	40,887	P-GFI
1973	43,118	11,099	13,654	46,156	Lachowski et al.
1976	33,147	6,300	10,188	31,266	Bonita and Revilla
1980	28,730	7,800	6,630	34,650	FDC
1980	32,149	7,858	7,689	33,734	P-GFI
1987	28,555	7,410	7,444	27,637	SSC
1987	25,597	6,605	5,323	29,184	P-GFI

Table 7 demonstrates the continuous decline in forest cover since 1950. The two estimates that seem the most inconsistent are Lachowski et al. (1979) for 1973 and Eckholm (1976) for 1976. The shortcomings of Lachowski et al. (1979) have been discussed above. Eckholm's estimate is clearly incompatible with the others, but since no reference is given it was included only for the sake of completeness. Table 8 presents forest-cover data at a superregional level for those years for which it is available. It is derived from table 7.

On the basis of tables 7 and 8, rates of change in forest cover (absolute and percentage) can be calculated at the national and superregional levels. The percentage rate of change is the average annual rate of change between two dates and is determined by dividing the change in forest cover by the average of forest cover at the beginning and end years. Tables 9 and 10 present the national and superregional rates of deforestation respectively. Obviously, since not all of the data are consistent over time, especially at the superregional level, the choice of beginning and end years and the choice of survey will influence the rates of change. Thus tables 9 and 10 should be interpreted with the understanding that the rates of change presented are only for the years and surveys chosen.

Table 9.
National deforestation rates in the Philippines, 1950–87

Years	Average annual change (km^2)	(%)	Source
1950–57	2,210	1.58	Projection and NEC
1957–69	2,262	1.91	NEC and P-GFI
1969–76	2,081	2.14	P-GFI; Bonita and Revilla
1976–80	3,048	3.64	Bonita and Revilla; FDC
1980–87	1,570	2.17	FDC and P-GFI
1950–69	2,243	1.78	Projection and P-GFI
1969–87	2,103	2.46	P-GFI and P-GFI
1950–87	2,175	2.03	Projection and P-GFI

Source: Table 7.

Table 10.
Superregional deforestation rates, 1950–87

Years	Annual rates in km^2 (and %)			
	Mindanao	Palawan	Visayas	Luzon
1950–69	1,007 (1.9)	46 (0.5)	350 (2.4)	839 (1.7)
1969–80	1,246 (3.5)	184 (2.1)	435 (4.8)	567 (1.5)
1980–87	448 (1.7)	171 (2.4)	187 (3.1)	781 (2.5)
1950–87	972 (2.2)	111 (1.3)	344 (2.9)	747 (1.7)
Total deforested (1950–87) (km^2)	35,979	4,098	12,743	27,652
Percentage of total forest loss (1950–87)	44.71	5.09	15.84	34.36

Source: Table 8.

The data presented in tables 7 through 10 demonstrate that forest cover has been declining continuously since 1950 throughout the Philippines. Deforestation was more rapid from 1969 to 1987 than from 1950 to 1969, with the highest rates occurring from 1976 to 1980. A surprising result of the above analysis is that the Visayas region has had the highest rate of deforestation since 1950. This seems to contradict the notion of Mindanao as the premier destination for rural migrants and the leading area for commercial logging. At the same time, Mindanao accounted for 45% and Mindanao and Luzon 79% of all deforestation in the postwar period.

A disturbing feature of the data presented in tables 7 and 8 is that rates of deforestation will differ depending upon the inventory chosen. Table 11 demonstrates how rates of deforestation can vary according to the different data sets. The 1980 forest data are from the FDC (manual interpretation of a LANDSAT photomosaic) and the P-GFI (projected back from 1987). The 1987 data are from the SSC and the P-GFI.

Needless to say, the differences among the four possible combinations of the two surveys each for 1980 and 1987 are great. At the superregional, regional, and provincial levels, the discrepancies are even greater. All agree that deforestation is occurring, but the difference between the smallest and largest rates of deforestation is more than 200%. The differences in the inventories and, more important, in the calculated deforestation rates, raise very perplexing questions with regard to the historical reconstruction of forest cover, the

Table 11.
Annual rates of deforestation between 1980 and 1987 for different national forest inventories

1980 Data	1987 Data	Km^2	Percentage
FDC	P-GFI	1,571	2.17
FDC	SSC	951	1.28
P-GFI	P-GFI	2,103	2.84
P-GFI	SSC	1,483	1.95

Source: Table 7.

Table 12.
Government statistics on national forest cover vs. inventory statistics, 1969–87 (percentage of total land area)

Year	Table 7		Government statistics
1969	P-GFI	34.9	53.6
1973	Lachowski et al.	38.0	46.3
1974	Bruce	29.8	45.6
1976	Bonita and Revilla	30.0	44.2
1980	P-GFI	27.1	41.5
1987	SSC	23.7	30.6[a]
1987	P-GFI	22.2	30.6[a]

Sources: Table 7; various issues of PFS.
[a]1986.

projection of future deforestation rates, and the role that tropical deforestation has played in contributing to the buildup of global atmospheric CO_2.

Philippine Government Statistics

The history of the decline of forest cover in the Philippines, as presented above, was based primarily on remotely sensed data, with the exception of the 1957 NEC data, which were based on BF field reports. While I have already noted several occasions where inventory results and official government statements have disagreed with regard to the extent of forest cover, Table 12 compares the statistics used in this study with the statistics presented in the PFS for the same year.

Forest cover reported in the PFS between 1980 and 1986 drops sharply because the results of the P-GFI became available for certain of the regions; in fact, for 1987 the results of the P-GFI and the PFS are almost the same. However, the discrepancy between the two sets of data since 1969 is substantial. In addition, it would appear logical that if forest cover is overstated, then the rate of deforestation will be correspondingly understated, as is demonstrated in table 13.

The figures presented in the 1987 PFS strongly disagree with the calculations of this study. For the 1970–87 period, I calculate that deforestation was more than 250% greater than the official statistics indicate for the same period. In fact, the government figures for most of the 1980s are so low as to have virtually no credibility. It is not my primary objective to engage in an extended analysis of official Philippine forestry statistics. I simply note that they have often ignored or contradicted other sources regarding the extent of forest cover and rate of deforestation. As the World Bank (1989a, 10) points out, there has been a "reluctance to acknowledge the extent of deforestation which has occurred." These data problems are further compounded by the failure of government reports on forestry to present their data in a way which can be examined critically.

Table 13.
Rates of deforestation: Government estimates vs. table 9, 1970–87 (km^2)

Date	Table 9	Government statistics
1970	2,081	1,700
1971	2,081	1,700
1972	2,081	1,700
1973	2,081	1,400
1974	2,081	1,400
1975	2,081	1,400
1976	2,081	840
1977	3,048	840
1978	3,048	660
1979	3,048	622
1980	3,048	326
1981	1,570	246
1982	1,570	167
1983	1,570	1,213
1984	1,570	49
1985	1,570	146
1986	1,570	77
1987	1,570	71
Annual average, 1970–87	2,097	787

Sources: 1987 *PFS* (table 1.13); Table 9.

Some of the difficulties involved in interpreting forest-cover data are demonstrated by the information contained in volume 7 (*Forestry*) of the eight-volume set on Philippine natural resources published by the NRMC (1979). On page 14 of this volume, the authors note that forest cover in 1979 totaled 12,864,326 ha, or 42.88% of total land area. On page 19, the authors discuss the results of the LANDSAT-assisted forestry inventory conducted jointly by the BFD and the General Electric Co.. They suggest that the LANDSAT inventory showed forest cover of 11,402,664 ha (38.01%) in 1976, which is incorrect because most of the imagery is from 1972–73, and 1973 is the proper midyear date. The authors then go on to compare the BFD statistics with the LANDSAT inventory results. "The comparison of the outcome of this forest inventory with the figures from the Bureau of Forest Development indicate that the results are not very widely apart. Taking the land use data of 1974 . . . the total area of actual forests [is] 13,690,058 [ha]. Thus, 45.6 percent of the total land area of the country is estimated to be covered by actual forest" (p. 19).

In short, the authors of this report made the following claims: forest cover was 45.6% in 1974, 38.01% in 1976, and 42.88% in 1979. Given that the results are inconsistent and the date of the LANDSAT imagery is incorrect, this report is difficult to interpret and does not further our understanding of forest cover in the Philippines.

Table 14.
Vegetative cover in Ilocos Norte, 1967 and 1980 (ha)

	1967	1980	Change
Open and cultivated	44,616	63,980	+ 19,364
Brushland	48,152	40,808	− 7,344
Dipterocarp (primary)	23,076	52,818	+ 29,742
Dipterocarp (secondary)	36,580	20,348	− 16,232
Pine (primary)	—	2,231	+ 2,231
Pine (secondary)	—	1,025	+ 1,025
Mossy	18,655	13,854	− 4,801

Source: Salvador et al. (1985).

Another example of the government's questionable interpretation of forest-cover data is contained in Salvador et al. (1985). Their study is an inventory of forest cover on public land in Ilocos Norte (Luzon) using aerial photos taken in 1980 compared with 1967 photos taken in the first national forest inventory (1962–68). Table 14 presents their interpretation of the two sets of photos.

Of particular importance is the following statement by the authors: "The increase of 31,973 ha of primary forest [29,742 in primary dipterocarp and 2,231 in primary pine] must have come from the young growth of the past inventory which was about 36,580 ha. It is assumed that after 13 years the young growth forest have grown to maturity. Thus, secondary forest decreased correspondingly. The increase rather than decrease can be attributed to the effect of the Presidential ban on logging in the province" (p. 6). And they conclude (on p. 11) that the 19,363 hectare increase in open and cultivated areas indicates an "alarming" rate of forest denudation of 1,210 ha per year.

The following comments and questions would seem to be appropriate:

1. Since total forest area increased from 78,311 ha in 1967 to 90,276 ha in 1980, it is not clear how this demonstrates that forest denudation is a problem.

2. Where did the primary and secondary pine forest come from?

3. Since dipterocarps take decades to mature, how could the primary dipterocarp forest increase by 29,742 ha in just thirteen years? As Segura et al. (1977) point out, according to the BFD's own sustained-yield management concepts it takes at least thirty-five years for residual forests to grow to harvestable stands.

4. The increase in primary dipterocarp forest was 29,742 ha, and yet secondary dipterocarp forest declined by only 16,232 ha. Where did the other 13,510 ha of primary dipterocarp forest come from?

5. Total hectarage is 171,079 in 1967 and 195,064 in 1980. Where did the increase of 23,985 come from?

Needless to say, numerous other questions could be raised, but the ones listed above are the most obvious. It should also be pointed out that the P-GFI results for this same period indicate a decrease in forest cover in Ilocos Norte and an increase in deforestation in those provinces or regions that had logging bans (Schade 1988). In short, the work of Salvador et al. (1985) raises more questions than it answers.

There has often been disagreement among non-Filipino sources regarding forest cover and rates of deforestation. As an example, the ADB (1987a) estimated that deforestation was occurring at the rate of 6,400 km^2 a year between 1968 and 1979, and Myers (1980) estimated a deforestation rate of 5,000–6,000 km^2 for 1980. On the other hand, FAO/UNEP (1981) calculated a deforestation rate of 1,010 km^2 per year for the 1976–80 period and projected deforestation for the 1981–85 period at 910 km^2 a year.

Since the FAO/UNEP study (1981) has been so extensively quoted, it is of interest to compare their results with my own. The FAO/UNEP (1981) projected 1970 BFD data to arrive at a total figure for forest cover of 130,200 km^2 for 1980. This is equivalent to 43.4% of total land area. If forest fallow (35,200 km^2) is not included, then forest area is 31.1% of total land area. By way of contrast, I have arrived at a figure of between 25.9 and 27.1% of total land area covered by forests in 1980 (see table 5). In addition, the FAO/UNEP rates of deforestation for 1976–80 and projected for 1981–85 are considerably below the rates of deforestation determined by my study. The differences are substantial and could have profound implications for ongoing efforts in modeling global warming and the global carbon cycle.

Given the wide range of estimates for forest cover and rates of deforestation in the Philippines from 1950 to 1987 that have been reported in the literature, I accept the figures presented in tables 9, 10, and 11 as the most reliable available. At the same time, I acknowledge that the data presented are subject to shortcomings and should be considered with these limitations in mind.

The Causes of Deforestation: Discussions in the Postwar Philippines

The material presented above indicates that official statistics regarding Philippine forest cover have often been incorrect and misleading. As a result, for much of the postwar period, it has been difficult, if not impossible, to obtain a clear view of the extent of forest cover or the rate of deforestation. Under these circumstances, it is not surprising that discussions about the causes of deforestation in the Philippines have often been characterized by statements which can be neither proven nor disproven.

Almost all observers agree that the two major causes of deforestation have been logging (legal and illegal) and the expansion of agriculture (Boado 1988; Nelson 1984; U.S. Department of State 1980). However, owing to the synergism between the two, it is difficult to assign primacy to any one cause

(Johnson and Alcorn 1989). The major question raised in the literature appears to be: who is to blame the most for deforestation, loggers or agriculturists? And, if agriculturists are to blame, to what extent are shifting cultivators the culprit? In short, is it possible to rank the causes of deforestation according to their relative contribution to forest destruction? A clear-cut answer to this question would facilitate devising policy measures to slow down or reverse the process of deforestation.

There are four major reasons why these questions are difficult to answer. First, since "shifting cultivation" and *kaingin* are rarely defined, there is a great deal of ambiguity surrounding the use of these terms. We simply note that most participants in discussions regarding deforestation in the Philippines either implicitly or explicitly equate *kaingin* with shifting cultivation, a point of view not accepted in this study. Second, it may be the case that the ultimate cause of migration to the uplands and deforestation is the failure of macroeconomic policies to provide employment and eliminate poverty in the lowlands. If so, deforestation is the end result of the failure of development to raise the standard of living for the majority of people, compounded by the open-access nature of the Philippine forests. Third, there are very severe data problems, and without information on a large number of variables at the local level, it may be difficult to determine the causes of deforestation. Fourth, the process of commercial logging and expansion of agriculture (permanent, shifting, or both) appear to go hand in hand. Numerous commentators have stated that logging facilitates agriculture by building roads and opening up the forest (ADB 1976; Boado 1988; Capistrano and Fujisaka 1984; Edgerton 1983; Guppy 1984; Marx 1985; Myers 1980; Pantastico et al. 1982; Porter and Ganapin 1988; Spencer 1966; World Bank 1989a). Consequently, in areas where agriculture has followed commercial logging, it is difficult to identify a single cause of deforestation.

If logging does precede agriculture, it should be an easy matter to verify statistically; unfortunately, this is not the case for several reasons. First, illegal logging has been widespread but there are no data on where and when it has occurred. Second, no comprehensive studies have yet been conducted regarding agriculture on formerly forested lands held by logging concessions (Cornista et al. 1986). Third, the poor records of the FMB, and possibly even poorer records of the logging companies, most likely preclude a detailed reconstruction of the history of logging within concessions (Coppin and Lennertz 1986; Willet 1976). As a result, longitudinal studies of the relationship between logging and the spread of agriculture may be difficult. However, the data problems should not obscure the fact that, in general, agriculture has expanded by occupying recently logged forests (Hicks and McNicoll 1971).

Although no studies of the expansion of agriculture onto previously forested or logged-over areas have been conducted, Uebelhor (1988) derived estimates of residual forests as a function of time elapsed after logging. The figures are tentative, since Uebelhor had to adjust FMB statistics that under-reported logging, which forced him to make broad assumptions regarding average timber volume per hectare for different time periods. Given these caveats, his calculations show that only 10% of the forests logged before 1955 still remain, while 95% of recently logged forests still remain. The main agent of the destruction of residual forests has been agriculture (both shifting and permanent). If these calculations are correct, it would indicate that there is a gap between logging and the spread of agriculture. In short, agriculture does not immediately follow logging. A major implication of Uebelhor's argument would have to be that logging company claims that farmers immediately follow logging operations are false. If Uebelhor is correct, it would appear that there is more room for proper protection of the forests on the part of concessionaires.

While numerous authors have made claims for the destructiveness of shifting cultivation (Gulcur [1968] claims that *kaingin* destroyed 172,000 ha of forest a year; the Philippine Council for Agriculture and Resources Research and Development [1982] claims 80,000 ha; Serevo et al. [1962] claim 2 million ha between 1937 and 1962; and Bee [1987] claims 330,000 ha a year between 1980 and 1985), I have yet to see any convincing evidence to support these assertions. It is obvious that forests are being depleted and that agriculture is increasing in area, but a detailed examination of the relationship between these two land uses over time remains to be conducted (Hyman 1983). Blaikie (1985) and Byron and Waugh (1988) also note that discussions of tropical deforestation often result in attempts to blame shifting cultivation or commercial logging and yet there is little hard evidence to support either position.

Since 1945, a group of foresters, primarily associated with the BF/FDB and private logging interests, has argued that commercial logging is not responsible for deforestation or is, at most, responsible for only a small part of it. Instead, they have claimed that deforestation is primarily caused by *kaingin* and illegal logging and, to a certain extent, the release of public lands for agriculture (Agaloos 1984; Generalao 1975; Llapitan 1983; Utleg 1970). As an example, the title of Agtani's article (1964) is "Kaidinism: Philippine Forestry's No. 1 Enemy." More recently, a paid advertisement in a major Manila daily blamed *kaingin* for 75% of all deforestation and illegal logging for the remaining 25%. In effect, the advertisement was stating that loggers were not responsible for deforestation (Lynch and Talbott 1988). The FAO (1987), in their study of forest resources in Southeast Asia, "assumes" that shifting cultivation is the "main agent" of deforestation and that logging in and of itself does not cause a "great deal of deforestation." The lack of data regarding the extent of forest cover, rates of deforestation, population in the uplands, and geographical expansion of agriculture has produced what Agaloos (1984) has called a "seemingly endless controversy" regarding who is to blame for deforestation in the Philippines.

More recently, several observers have emphasized that deforestation in the tropics is not simply a matter of forestry and agriculture; rather, political and social factors such as rapid population growth and unemployment and poverty in the lowlands are ultimately the deciding factors (Wyatt-Smith 1987). In the Philippines, the DENR et al. (1987, i) concluded, "The state of the environment and natural resources sector can be summed up as generally characterized by social inequities/injustices just like that of the nation in general." They note that poor people are being forced to use the forested uplands because of an unequal distribution of land and poverty in the lowlands, while at the same time the government has facilitated the use of natural resources by the rich and powerful.

Ganapin (1987) notes that while destructive logging, shifting cultivation, and population growth are the immediate causes of deforestation, the primary causes must be found elsewhere. He argues that the elitist society of the Philippines is characterized by concentration of wealth in the hands of the few, poverty for the majority of the people, and widespread corruption, and that it is this entire socioeconomic system which is conducive to forest destruction. The wealthy destroy the forests for private gain and the poor destroy the forests to feed themselves. These arguments are developed in more detail in Porter and Ganapin (1988).

Nelson and Cruz (1985) argue that the causes of deforestation and upland degradation are not to be found in the forestry sector at all. They note that deforestation is primarily caused by the migration of lowlanders onto recently logged lands and ask the question: "Why are alternative activities (e.g., in industry or lowland agriculture), which have fewer negative externalities, not as desirable?" (Nelson and Cruz 1985, 2). In short, why are forest resources being overexploited? Their answer is that the Philippine economy has not been growing at a rate sufficient to absorb excess labor and make upland agriculture an unattractive occupation. They claim that the major cause of this state of affairs is rapid population growth and low wages in the nonforestry sector. In short, macropolicies and the failure of development in the Philippines have made overexploitation of forest resources a rational activity.

The World Bank (1989a) rejects the line of reasoning of Nelson and Cruz (1985). They argue instead that the effects of macropolicies are trivial when compared to the large economic rents that are available to forest exploiters (loggers, agriculturists, commercial charcoal makers). At the same time, they note that poverty in the lowlands is the result of "excessive" population and the inequality of income and assets and that this is causing the migration of poor people to the uplands.

The works just reviewed indicate that the context in which deforestation occurs must be considered to be as important as a study of the main instruments of deforestation, whether they be loggers or agriculturists. At the same time, this is not to argue that logging and agriculture are the only two causes of deforestation in the Philippines. For instance, commercial charcoal making has been related to deforestation in the Philippines (Robert Huke 1986; personal observation of author) and mangroves have declined drastically in the twentieth century, with conversion to fish and salt ponds and human settlements being the primary reasons. Thus, even though it would appear that the two major immediate causes of deforestation in the Philippines are the activities of loggers and agriculturists, I recognize that there are exceptions to this statement. However, these exceptions are not considered to be important in relation to total forest cover.

I have already noted on several occasions the difficulties of working with data from the FMB. Nowhere is this more apparent than in attempting to analyze the causes of deforestation. The FMB has either conducted or been involved with five national forest inventories since 1962 and very little attempt has been made to analyze the data in such a way that it could contribute to discussions surrounding the issue of deforestation (Nilsson et al. 1978 made the same observation). The one major exception to this are the regional reports produced by the P-GFI. These reports provide detailed information on rates of deforestation at the provincial level, timber volume, and types of forest cover. While the data of the P-GFI project do not allow a calculation of the percentage of deforestation that is the result of any one particular cause, the regional studies state that the major causes of deforestation have been logging, shifting cultivation, and the expansion of agriculture.

Unfortunately, the other inventories are not so helpful. As mentioned above, the second national forest inventory (1965–72) appears to have been suppressed. The only results of that inventory are a few pages in the 1973 *PFS* and, as far as I am aware, no interpretation or analysis of the results was ever conducted by the BFD. The LANDSAT-assisted forest inventory (General Electric and NMRC 1977; Lachowski et al. 1979) also did not discuss deforestation. No comparison was made to the results of the 1962–68 and 1965–72 inventories and thus rates of deforestation were not calculated. In addition, no nationwide estimates of agricultural lands or rangelands were made.

As mentioned above, the results of the first national forest inventory (1962–68) are available only for Palawan and Mindanao. The results for Palawan cover the period 1948–64 and for Mindanao, the period 1952–63. Since these results are the most comprehensive of any government studies done in the postwar period with the exception of the P-GFI, table 15 presents them.

For Mindanao as a whole, Agaloos (1964a, 1964b, 1965) notes that forest area declined by 1,007,000 ha between 1952 and 1963. He also notes that of the 20.1 million m^3 of timber lost during this period, 4.8 million m^3, or 24%, was accounted for by commercial logging. In other words, approximately 75% of the timber drain was the result of the release of lands for agriculture and unauthorized clearing. At the same time, the decline in the extent of the old-growth dipterocarp forest is in large part the result of commercial logging. For

Table 15.
Average annual land-use change in Palawan (1948–64) and Mindanao (1952–63) (in ha)

	Palawan	Mindanao
FORESTLAND		
Productive forest		
Reproduction brush	+ 3,523	– 68,280
Young growth	+ 2,952	+ 49,787
Old growth	– 10,666	– 73,071
Total	– 4,191	– 91,564
Unproductive forest		
Total	0	0
All forestland	– 4,191	– 91,564
NONFORESTLAND		
Crops and pasture	+ 1,928	+ 65,414
Plantation crops	+ 816	+ 34,078
Open land	+ 1,437	– 10,540
Marshlands	0	– 310
Urban and other	+ 10	+ 2,922
All nonforestland	+ 4,191	+ 91,564

Sources: Agaloos (1964a, table 5); Agaloos (1964b, table 5); Agaloos (1965a, table 5); Agaloos (1965b); Agaloos and Santos (1968, table 3).

Palawan, Agaloos and Santos (1968) argue that virtually all of the loss of forestland has been a result of agricultural development because, in their view, commercial logging was almost nonexistent before 1964.

When interpreting the results of the first national forest inventory, it is important to remember the following: first, the results for the Visayas and Luzon were never published; second, the reported timber drain caused by commercial logging in Mindanao and Palawan must be considered an under-estimate since it does not include illegal logging and overcutting of AACs by concessionaires; third, the period covered occurred before the dramatic increase in commercial logging in the 1960s and 1970s.

Another concern has to do with the determination of forest cover and timber volumes. The P-GFI uncovered numerous instances where forest cover on the 1969 FRCMs was incorrectly labeled and it is possible that the same may have happened with the interpretation of the 1960s aerial photography. In addition, there is the possibility that timber volumes could have been overestimated if the sampling procedure was biased toward lowland dipterocarp forests, where timber volumes per hectare are much greater than at higher elevations.

Even if we assume that forest cover has been correctly identified and timber volumes are accurate, there is still a problem with the definition of forestlands used in the studies of Palawan and Mindanao. The project defined forestlands as lands that are at least 10% stocked with trees; the implications of this definition will be discussed below. As defined by the BF, unproductive forests are mossy or cloud forests found at higher elevations and have no commercial value. The inventory results for both Palawan and Mindanao indicate no change in the extent of these forests. Productive forests include mangrove and dipterocarp forests and the following definitions used in the study are from Agaloos (1964a):

> Reproduction brush: "Productive forest stands predominantly stocked with tree reproduction or brush. Trees at least one meter high but less than 15 cms. in diameter are classified as reproduction."

> Young growth: "Productive forest lands predominantly stocked with trees 15 cms. or larger in diameter. Most trees in this class have been cut over, with residual trees remaining. Stands stocked mainly with mature trees but with 25 percent or more of the mature stand volume removed by cutting, qualify as young growth."

> Old growth: "Productive forest lands predominantly stocked with mature trees with less than 25 percent of the mature stand volume removed by cutting."

The combination of the definition of forestlands (lands stocked at least 10% with trees) with the definition of reproduction brush can easily lead to a mental image of a grassland with a small group of saplings over 1 m tall being classified as forestland. In short, I find the definition of forestland to be rather weak. The fact that the 10% limit was also used by FAO consultants in the late 1970s in the Philippines (Nilsson et al. 1978; Singh 1979) can easily lead to the speculation that the definition had been chosen to inflate the extent of forestlands. This definition must be considered to be a major weakness of the 1962–68 forest inventory.

The definition of young-growth forest makes it clear that these were originally old-growth forests that have been logged at least once. Thus, the annual increase in young-growth forests of 49,787 ha in Mindanao most likely comes primarily from old-growth forest. This would certainly be the case if selective logging had been practiced; although this study can not eliminate the possibility that clear-cutting was practiced in some areas.

Based on the analysis above, 68% of the decrease in old-growth forests (49,787 out of 73,071 ha per year) is the result of logging, and the remaining decline in old-growth forest is going either directly to cropland or to reproduction brush. The latter possibility is doubtful, since this category itself decreased substantially. According to Agaloos (1964a, 1964b, 1965b), reproduction brush is declining because these areas are being converted to cropland and the percentage of land in reproduction brush that becomes young growth is quite small.

Of the timber drain of 20.1 million m^3, 12 million m^3 came from merchantable dipterocarps (55 cm or larger) and, of this, 36% (4.3 million m^3) was caused by logging. Once again, it is important to remember that this is authorized logging only. If we assume underreporting within concessions and no reporting for illegal logging outside concessions, then this percentage would obviously be higher. In fact, it would seem to support the reasoning above that the conversion of old-growth to young-growth dipterocarp forests has primarily been accomplished by logging.

In general, the pattern of land-use change in all three areas of Mindanao is practically the same: a decrease in reproduction brush, old-growth forests, and open land and an increase in young-growth forests and cropland. Given the data though, a detailed description of the agriculture/forest interaction is not possible.

One consequence of this inventory is that it appeared to exonerate loggers of being a major cause of deforestation. Since logging was responsible for less than 25% of the timber drain, the logical conclusion was that slightly more than 75% was caused by agriculture of some sort (Hicks and McNicoll 1972). However, I do not accept this interpretation for several reasons.

First, the definitions of forestlands and reproduction brush are such that large areas of grasslands could conceivably be included within the forest category. Second, the data for authorized logging most likely understate the total amount of timber loss caused by logging by a substantial margin. Third, a logical analysis of the land-use categories and land-use changes indicates that most of the loss of old-growth forest came from logging. In short, while I agree with the conclusion of the first national forest inventory that forest loss was serious, I cannot accept its emphasis on agriculture to the exclusion of logging as the cause of deforestation. A closer examination of the data on deforestation in the inventory indicates that logging was one of the major causes of the decline in the primary forest of Mindanao.

My interpretation of data from the first national forest inventory suggests that logging played a crucial role in the transformation of primary forest to secondary or residual forest. This interpretation receives strong support from data at the national level. Schade (1988) notes that between 1969 and 1988, an average of 1,800 km² of virgin forest were logged each year. At the same time, agriculture (shifting and permanent) expanded by 1,900 km² a year: primarily through conversion of residual forests. The World Bank (1989a) points out that old-growth forests have declined from approximately 10 million ha in the 1950s to approximately 1 million ha today and that the main cause of this was logging. In addition, Marx (1985) documents the same process in the province of Quirino on the island of Luzon.

In short, even though the discussions in the postwar Philippines regarding the relative effects of logging and spreading agriculture on deforestation have been inconclusive, there is evidence to indicate that the destruction of the primary/virgin/old-growth forests has primarily been accomplished by logging (legal and illegal). At the same time, the destruction of the residual forests has come about mainly at the hands of agriculturists. While Uebel-hor's work (1988) indicated that there may be a gap between logging and the expansion of agriculture, it is only suggestive, and the lack of microstudies precludes a definitive statement in this matter.

Corruption in the Forestry Sector

Illegal activity in the forestry sector takes many forms in the Philippines. For purposes of this study, the three major types of illegal activity are: (1) illegal logging, which is logging for commercial purposes without a permit or exceeding the ACC if one has a permit or concession; (2) the underinvoicing of timber exports and smuggling of logs overseas; and (3) the granting of concessions either for political or monetary favors. Many of the more common forms of corruption, such as payments to military personnel at checkpoints and bribes to forestry officials, stem from activity which is illegal as defined above. The occupation of public forestland and upland areas by *kaingineros* and native peoples is not considered to be illegal in this discussion.

The concept of illegal activity can be extended to include improper logging techniques, poor construction of logging roads, failure to reforest after logging, and, perhaps most important, failure to prevent agriculturists from occupying concession land which has been logged. All of these activities can contribute to deforestation by facilitating the spread of agriculture and increasing the rate of logging; however, hard evidence to substantiate the above claims is limited. Therefore, the remainder of this chapter will be devoted to a wide-ranging discussion of corruption and related issues which will approach the subject from several different viewpoints. The goal is to come to a better appreciation of corruption and its relationship to deforestation in the postwar Philippines.

For large-scale corruption to have occurred in the Philippine forestry sector, four conditions had to be met. First, a market had to exist for Philippine forest products. This role was played primarily by the domestic market right after World War II that responded to the demand for wood for reconstruction and by Japan and the United States after 1950. Second, a supply of tropical timber had to exist, and this the rich Philippine forests easily provided. Third, capital-intensive equipment for large-scale commercial operations had to exist. While this was not the case in 1945, by the early 1950s the Philippines had the most capital-intensive forestry sector in Southeast Asia, if not the entire tropical world (Edgerton 1953; Philippine Council for Agriculture and Resources Research and Development 1982). Fourth, the sale of tropical wood products had to generate a sufficient financial return to encourage logging. While data on the actual profits made by loggers in the Philippines or other tropical countries are scarce, the evidence reported by Repetto (1988) and Boado (1988) demonstrates the existence of substantial economic rents; that is, economic returns in excess of the cost of production and normal profits. In short, the initial conditions for large-scale commercial logging and exports have existed in the Philippines since shortly after independence was granted in 1946.

That corruption on a massive scale has been common in the Philippines, particularly under the Marcos leadership (1965–86), is generally accepted (Alano 1984; Aquino 1987; Clark 1988; Hainsworth 1979; Magno 1985; Mydans 1988; Tapales 1986; Wurfel 1979). To a certain extent, the entire economy has been manipulated for the benefit of a small number of families, and corruption in the sugar and coconut industries was particularly blatant under former president Marcos (Carbonell-Catilo 1986; DeDios 1984; Hill and Jayasuriya 1984). Some idea of the magnitude of corruption in the Philippines can be garnered from the Commission on Audit's estimate that for the 1975–80 period, 10% of the entire GNP went into bribes (Tapales 1986); Alano (1984) has calculated that for the 1965–78 period, 29–54% of the entire export value of goods shipped to the Philippines was subject to some form of technical smuggling. Facilitating this sort of widespread corruption was the fact that by the mid-1970s, 65% of all government positions were filled without competitive exams; that is, they were political appointments (Oshima 1987).

It should come as no surprise, therefore, that corruption and private use of public office are accepted as normal in the Philippines (Carino 1986; Palmier 1989; Shahani 1988; Stone 1971). Stone (1971) has examined the relationship between bribe-taking policemen in Manila and the street vendors and jeepney drivers who pay the bribes. He concluded that public space, because it is owned by no one, is actually under the temporary control of the government agent on the scene; in this case, the policeman. Thus, bribes are offered (and accepted) for the short-term use of public space. This analysis can be extended to other areas of the Philippine economy, and forests are most likely the premier example of a public resource being used for private profit (Porter and Ganapin 1988). That competition for this resource was intense was indicated by Poblacion (1959, 90) when he noted that it was becoming difficult for prospective loggers to find virgin forest, since "almost all the accessible commercial forest areas are now occupied." It should also be kept in mind that when Poblacion spoke concessions totaled about 40,000 km^2, and eighteen years later, when forest cover was much less, concessions covered 102,000 km^2. Bello (1988) has argued that the Philippine political system is basically a form of "institutionalized looting," with the main purpose of public office being to enrich oneself and one's followers—a characterization I would accept as correct.

Private use of public forests has not escaped the attention of journalists. Numerous articles have appeared in the press regarding illegal logging (Flores 1986; Robles 1987), and the *Philippine Lumberman*, which primarily represents the private forestry sector, has run literally hundreds of articles and news clips on log smuggling, corruption, the use of political influence to acquire concessions, and the incompetence of the BF/BFD staff. The topic has not suffered from lack of discussion. However, an outstanding characteristic of Philippine corruption is that virtually no one seems to be punished for such activities (Palmier 1989). The World Bank (1989a, 5) in a carefully guarded statement regarding industrial polluters, notes that "governmental sanctions of any kind are not notably successful in the Philippines." This would certainly seem to be the case in forestry. Cancellation of concession licenses seems to be the harshest form of punishment in the forestry sector; however, Dugan (1987) reports a case where loggers who were engaged in large-scale smuggling and had their licenses canceled were able to get them re-instated by a high-ranking government official in partnership with them. As another example, in 1985, the Governor of Lanao del Sur was caught export-ing illegally cut logs but was protected by then-president Marcos (Doeppers 1986).

Figures on the extent of illegal logging do not exist; however, an idea of the magnitude of the problem can be gathered by comparing official Philip-pine exports of logs with official imports of Philippine logs by its trading part-ners. Durst (1985), using data from 1976 to 1980; Sanchez (1986), using data from 1977 to 1985; Power and Tumaneng (1983), using data from 1977 to 1980; and Pernikar (1984), using data from 1978 to 1982, present evidence on the magnitude of this discrepancy. As an example, Sanchez (1986) notes that Japanese log imports from 1980 to 1982 were approximately 250% greater than official Philippine log exports to Japan. Although Japan has taken by far the largest percentage of Philippine forestry exports, the discrepancies occur with other countries, as Durst (1985) and Pernikar (1984) document. Table 16 shows the effect of including illegal exports and illegal logging for the domestic mar-ket on Philippine forestry production statistics.

Pernikar's calculations indicate that the discrepancy between official and unofficial logging data was growing over the 1978–82 period, and both Durst (1985) and Sanchez (1986) report the same phenomenon. Even the Na-tional Environmental Protection Council (1983) reports the same for 1978–81. And if the *Economist* (1989) is correct, the problem has become worse: it re-ports that official timber exports in 1988 were U.S. $200 million but that the actual figure is most likely over U.S. $1 billion. Unfortunately, data for earlier

Table 16.
Revised log production, 1978–82 (in 1,000s m³)

	BFD volume	Illegal volume	Revised volume
1978	7,169	500	7,669
1979	6,578	1,000	7,578
1980	6,352	1,500	7,852
1981	5,400	2,000	7,600
1982	4,133	3,000	7,133

Source: Pernikar (1984, 39, table 9).

periods do not exist, but the fact that the Philippine Central Bank in 1964 was already questioning the reliability of log export statistics indicates that this is not a new phenomenon (Palo 1980).

It is now generally accepted that the commercial forest resources of the Philippines have been underpriced throughout the postwar period (Angeles 1982; Boado 1988; Cruz and Angeles 1984; Power and Tumaneng 1983; Repetto 1988). Repetto (1988) argues that the main reason for the rapid increase in logging in the postwar period was the large profit that accrued to those who held concessions. He estimates that between 1979 and 1982, for example, the government captured only 12% of the available economic rents of U.S. $1 billion. In other words, "more than US$ 820 million remained with private timbering interests and their allies as abnormal profits" (p. 61).

However, data on actual profits or the amounts paid in bribes are almost nonexistent. Flores (1986) reports that logging trucks coming from the Cagayan Valley in northeast Luzon pay 7.5 million pesos a month to military checkpoints between Cagayan and Manila. (One U.S. dollar is worth approximately twenty-seven pesos.) Since this figure does not include payments to policemen, FMB inspectors, or politicians, it must be considered a conservative estimate.

A major incentive to smuggle logs abroad is that logs not reported are free of any export tax. However, an even greater incentive may be that nonreporting of log exports means that foreign exchange derived from the sale of these logs can be hidden abroad. Once again, the extent of this activity is almost impossible to determine, although the following example may give some idea of the possible scale. When Indonesia opened up its forest to commercial exploitation after former president Sukarno's ouster in 1966, Philippine logging firms were among the first to start operations. By 1969, ten Filipino firms had invested U.S. $275 million in forestry projects in Indonesia (*Philippine Lumberman* 1969). It was widely believed by officials of the Philippine government that much of this money came from sources which had been illegally salted abroad (*Philippine Lumberman* 1970). This same practice has also been reported to have occurred during the 1980s (*Philippine Lumberman* 1985). Salting of foreign exchange abroad is not confined to the forestry sector; Andal (1986) reports that in the garment industry alone, approximately U.S. $370 million was salted abroad between 1965 and 1983.

The salting of dollars abroad raises an interesting question regarding the financial profitability of forestry operations. If a currency is overvalued (as the Philippine peso has been for almost the entire postwar period) or if the desire to hold foreign exchange is strong, then an evaluation of the profitability of a logging operation solely in terms of pesos may not be the relevant measure of its viability to the owner. It is conceivable that a logging operation which is only marginally profitable or actually losing money in peso terms could still represent a substantial income stream to its owners if export receipts were underreported or not repatriated. Given the relative lack of investment opportunities in the postwar Philippines, the incentive to salt dollars (or yen) abroad may have been great.

To the above discussion of corruption in the Philippine forestry sector I would add my personal observations. I have viewed drivers of logging trucks paying soldiers at checkpoints, small-scale commercial logging within a national park, illegal charcoal making on a commercial scale within a reforestation site run by the FMB, and large-scale commercial logging on a 100,000 ha concession which was supposedly shut down by the government. In the latter case, logging trucks were passing within yards of a Philippine army outpost on a daily basis.

In addition, personal communication with dozens of people who are knowledgeable about forestry in the Philippines (and who wish to remain anonymous) has confirmed the general picture presented above; corruption and illegal activity are the norm and have been for decades. Even the supposedly more professionally run forestry companies are not immune to illegal activity: for years one of the largest forestry firms in the Philippines was exceeding its ACC by at least 20% so as to increase its foreign-exchange earnings (personal communication, anonymous source). Large-scale corruption in the forestry sector has also been observed in Thailand, Malaysia, Indonesia, and Western Africa. In fact, corruption has occurred wherever logging has been practiced on a commercial scale in the tropical world. Whether or not corruption is more or less widespread in the Philippine forestry sector as compared to Indonesia or Malaysia is irrelevant for our purposes and impossible to prove either way. That it has occurred on such a large scale means that official statistics on logging may be inadequate to truly represent the extent of logging activity.

I accept the fact that illegal activity as defined above has been the norm in Philippine forestry. In this light, the official forestry statistics discussed previously make more sense. The misrepresentation of forestry data either through omission or commission has been part and parcel of the corruption which has pervaded Philippine forestry. Much of the data presented by the BF/BFD regarding forest cover (at least up till 1986) has been a smoke screen behind which illegal activity has been able to hide. This point cannot be proven with any scientific certainty, but the circumstantial evidence makes it a reasonable proposition. The official forestry statistics for almost the entire postwar period have given a misleading picture of forest cover in the Philippines.

The contribution of corruption to the process of deforestation will be discussed in more detail below. At this point, we note that large-scale corruption facilitated the rapid expansion of legal logging, minimized efforts to stop illegal logging, tolerated destructive logging practices and ineffectual attempts at reforestation, and concentrated financial resources in the hands of loggers and their allies. I agree with the main thesis of Porter and Ganapin (1988, 13): "The Philippine political system has concentrated control of natural resources in the hands of the few at the expense of the economically disadvantaged and put a premium on the short-term exploitation of resources."

Summary

The evidence presented in this chapter indicates a drop in national forest cover from 70% to 50% during the period from 1900 to 1950. Since 1950, forest cover has dropped continuously and is now under 25% of land area. Approximately 55% of the forest cover in 1950 no longer exists.

A major question concerns the relationship between forestry and agriculture. Is the expansion of agriculture the cause of deforestation? Are deforested areas being converted to grasslands, permanent agriculture, or shifting cultivation? Chapter 4 will provide background information on the Philippines, including agriculture, which will help in the effort to put these questions in a more comprehensive framework.

4

The Postwar Philippines

Demography

Census data since World War II cover the years 1948, 1960, 1970, 1975, and 1980. Reliable population data after 1980 are not available. Table 17 presents the major demographic features of the Philippine population since 1948.

As table 17 indicates, the rate of growth of the Philippine population has declined slowly from the 1948–60 period to the 1975–80 period. However, it is presently the highest in Southeast Asia, and the Philippine population density is the highest of any country in Southeast Asia with the exception of Singapore (Population Reference Bureau 1990). In addition, the rural population as a percentage of total population has been declining slowly from 73% in 1948 to 62.7% in 1980.

Urbanization in the postwar Philippines, as in most of Asia, has been relatively slow (Jones 1983; Pernia 1988). The urban hierarchy is dominated by the primate city of Manila (Cressey 1960; Pernia 1988; Pernia et al. 1983); there

Table 17.
National demographic data, 1948–80

Year	Population (1,000s)	Average annual rate of increase (%)	Population density (persons/km^2)	Urban population (1,000s)	Rural population (1,000s)
1948	19,254	—	64.1	5,184	14,050
1960	27,085	3.06	90.3	8,072	19,015
1970	36,681	3.01	122.3	11,678	25,007
1975	42,070	2.79	140.2	14,047	28,024
1980	48,097	2.71	160.3	17,944	30,155

Sources: Population censuses, 1948, 1960, 1970, 1975, 1980.

is a distinct lack of intermediate-sized cities (National Economic and Development Authority [NEDA] 1982; Ullman 1960); and urban problems of poverty, squatting, pollution, and congestion have become major issues within the past twenty years (Abad 1981; Ramos-Jimenez et al. 1986).

Closely related to the phenomenon of urbanization is the history of migration in the postwar Philippines. Although it is now generally accepted that the Philippines has a serious population problem due to its size and rate of increase, this perception is fairly recent. Pelzer (1941, 1945) argued that the Philippines was not overpopulated; rather, since population was concentrated on a few islands, the major issue was the "maldistribution" of population. He suggested migration to the relatively underpopulated island of Mindanao as the solution. Duckham and Masefield (1969, 417) stated that the Philippines had a low population density and that there was "no real pressure of population on resources." As the discussion below will make clear, these comments seem almost naive today; however, the presence of areas with low population densities and available agricultural land has been an important factor in migration in the postwar Philippines.

Internal migration in the Philippines has been discussed by many authors (Abad 1981; Abejo 1985; Bernardo 1982–83; Bulatao 1976; Concepcion 1983; Fleiger et al. 1976; Institute of Population Studies 1981; Kim 1972; Krinks 1970, 1974, 1975; National Census and Statistics Office [NCSO] 1981; Nguiagain 1985; Perez 1985; Population Institute 1966; Pryor 1979; Simkins and Wernstedt 1971; Smith 1977; Ulack 1977; Wernstedt and Simkins 1965; Zosa-Feranil 1987). The major patterns since 1948 have been twofold: frontier migration primarily to Mindanao until 1960, and since 1960 a movement toward urban areas, particularly the Metro Manila Area. While migration to urban areas has been particularly pronounced since 1960, movement to frontier and upland areas is still continuing, as the work of Cruz et al. (1986) has demonstrated. In fact, between 1975 and 1980, almost one quarter of all interregional migrants had the uplands as their destination (Cruz and Zosa-Feranil 1988).

Overall, the major receiving areas have been Mindanao and the Metro Manila Area. The major pull factors of these areas have been available land and jobs respectively (Herrin 1985). The major sending areas have been the Visayas and the Bicol and Ilocos regions of Luzon. In general, migration has been to the more developed regions of the Philippines (Perez 1985) and migration has come more and more to be dominated by rural-to-urban and urban-to-urban flows as opposed to the rural-to-rural flows in the 1950s.

Table 18 shows population density by region for the postwar period and demonstrates that while substantial differences persist, population has become more evenly distributed since 1948 (Herrin 1985). In addition, the difference between regional and national population growth rates has dropped sharply from 1948–60 to 1975–80 (Cruz et al. 1986).

Table 18.
Population densities by region, 1948 and 1980 (persons per km^2)

Region	1948	1980
1 Ilocos	90	164
2 Cagayan Valley	21	61
3 Central Luzon	102	263
4 Southern Tagalog	56	130
5 Bicol	94	197
6 Western Visayas	125	224
7 Central Visayas	142	253
8 Eastern Visayas	82	131
9 Western Mindanao	41	135
10 Northern Mindanao	36	97
11 Southern Mindanao	15	106
12 Central Mindanao	29	97
Philippines	64	160

Source: Adapted from Zosa-Feranil (1987, table 2).

Economic Performance, 1948–88

By international standards, economic growth in the Philippines, at least until the early 1980s, has been above average. Annual percentage growth in GNP for the three decades since 1950 was 6.6 (1951–60), 5.1 (1960–70), and 6.2 (1970–81) and per capita income grew at 2.8% annually during the entire period (Crone 1986). While commendable, it should be noted that the economic performance of the Philippines was the lowest by far of all the ASEAN countries, with the next lowest, Indonesia, having a per capita growth rate of 4% annually between 1960 and 1980. In addition, events since 1983 (the year former senator Aquino was assassinated) have demonstrated just how fragile Philippine postwar development was. Percentage growth rates in GNP since then are as follows: 1.3 (1983), -7.1 (1984), -4.1 (1985), 2.0 (1986), 5.7 (1987), and 7.0 (1988) (Economist Intelligence Unit 1989).

Between 1983 and 1986, GNP declined by approximately 10%, and since population growth is between 2.4 and 2.8% a year, this means that per capita income fell by approximately one-fifth (the Asia Society [1986] reported a drop of 17% for the 1983–86 period). More important, this decrease in per capita GNP took place after a long-term decline in the living standards for the majority of people even when per capita income was increasing (to be discussed below).

The postwar reliance of the Philippines on an import substitution strategy and its shortcomings have been discussed by many (Hill and Jayasuriya 1984; International Labor Office [ILO] 1974; Oshima 1987; Pernia et al. 1983; Power and Sicat 1971; World Bank 1976). For our purposes, the major effects were that (1) import substitution was based on a plethora of rules and regulations regarding import and exchange controls and financial incentives which favored those with the proper contacts and led to corruption; (2) the overvalued exchange rate (necessary for cheap imported capital goods) had a negative effect on traditional agricultural exports; (3) high levels of protection had a negative impact on agriculture's terms of trade with the domestic manufacturing sector; (4) import substitution led to capital-intensive industries, located primarily in the Metro Manila Area, which were unable to absorb many of the migrants from the countryside; (5) while manufacturing grew at a rapid rate in the 1950s (particularly 1950–56), it started to decline in the second half of the 1960s and by the mid-1960s was no longer a leading sector; and (6) manufacturing failed to develop the backward and forward linkages necessary for self-sustaining growth and it has been able to provide only minimal support to the agricultural sector.

In short, the Philippines has not been able to develop a strong manufacturing base in the postwar period. The domestic manufacturing structure is still inward-looking and limited by the size of the market and lack of appropriate linkages with other sectors of the economy.

Oshima (1987) attributes the failure of Philippine development primarily to a "predatory" economic elite that developed under Spanish and American rule. Since the granting of independence in 1946, this oligarchy has refused to share power with other groups in society and has manipulated the political and economic structure for its own benefit. The well-documented excesses of the Marcos era (1965–86) are simply a more blatant example of what Oshima would consider to be normal for the "rapacious" Philippine oligarchy. While detailed information on the distribution of wealth and extent of elite control is almost nonexistent, Krinks (1983) suggests that sixty families exercise "significant control" over the Philippine economy.

Poverty and the Distribution of Income

In discussing poverty and the distribution of income, it is important to remember that from the late 1940s to the early 1980s, the Philippines experienced an almost continual increase in per capita income. Thus, we are talking about a country which according to contemporary and historical standards has been successful in achieving growth, at least until 1983.

The most comprehensive discussion of income and wealth distribution in the Philippines is Mangahas and Barros (1980). Their major findings are: (1) there are serious problems with much of the income and wealth data regarding coverage and accuracy; (2) given the data problems, seven national household surveys conducted between 1961 and 1975 indicated a Gini coefficient of between .44 and .64 with a mean of .53 (0.0 represents perfect equality and 1.0 represents perfect inequality); (3) the Gini coefficient appears to be fairly stable over time; (4) although the data on poverty are not entirely accurate, seven surveys based on food thresholds, income, and expenditures indicate that relative and absolute poverty worsened considerably between 1965 and 1975, and in 1975 it was estimated that at least 60% of the population was living in poverty; (5) available evidence indicates that the real wages of skilled and unskilled laborers have been declining since the mid-1950s and the share of income going to the propertied classes has increased.

The findings of Mangahas and Barros have been corroborated by several other authors. Lampman (1967) concluded that income distribution worsened from 1956 to 1961, and Wong and Arief (1984) and Khan (1977) report the same for 1956–71 and 1960–77 respectively. Hainsworth (1979) presents data which indicate that during the 1971–75 period the Philippines had a per capita caloric consumption less than that of India or Bangladesh. Hill and Jayasuriya (1984) report a decline in real wages from the 1950s to 1980, and Nelson (1984) claims that real wages have decreased since the 1960s. Panayotou (1983b) claims that there was no reduction in poverty in the 1960s and 1970s, and Crone (1986) and the Asia Society (1986) argue that poverty increased in the 1980s. The deteriorating condition of the mass of the Filipino people has not escaped the attention of international agencies, as the reports of the ILO (1974), World Bank (1976), and USAID (1980) indicate.

Lastly, poverty in rural areas is more severe than in urban areas, with average family income in the rural areas only 40% of urban family income (David 1987). Even more telling is the fact that as many as 95% of all school children in Bicol may be malnourished (Astillero 1976). Gorra (1986) estimates that between 20% and 30% of all schoolchildren and preschoolers are moderately to severely malnourished, a conclusion shared by the Republic of the Philippines/UNICEF (1987). Anderson (1982, 1987), Costello (1984), and Rocamora (1979) have all reached the conclusion that rural development has not improved the lives of the majority of the population and, in fact, in many instances, development has led to deterioration of living standards.

The poverty data reviewed above represent a negative assessment of Philippine development since independence and confirm the previous economic analysis. In fact, the outstanding characteristic of Philippine development has been its failure to raise living standards for the majority of its citizens (Oshima 1987). While the evidence is not completely reliable, it would appear that inequality of income and absolute poverty have increased since the late 1950s. Increases in per capita income during this period have flowed primarily to a relatively small group of people (Oshima 1987). The inability of the manufacturing sector to create a significant number of jobs in the urban areas means that the burden of job creation and poverty alleviation fell to the agricultural sector.

Agriculture and the Uplands

Agriculture has played and continues to play a major role in the Philippine economy. As the Agricultural Policy and Strategy Team (APST) (1986, 3) states:

The most glaring evidence of the failure of Philippine economic development is the fact that no significant structural transformation has
taken place over the past 25 years. Despite the strong industrial orienta-
tion of past economic policies, agriculture, fishery, and forestry continue to employ half of the labor force, contribute about a quarter of the
gross domestic production, and earn two-fifths of export revenues.
Over 60% of our population lives in the rural areas. Our country remains today as it has been in the past, a predominately rural society
composed of small farmers, agricultural laborers, fishermen, pedicab
drivers, and others.

David (1983) points out that agriculture's share of the total economy
declined slowly in the postwar period; from 36% of net value added in 1955 to
29% in 1980; and the World Bank (1989b) claims that agriculture's share of
gross domestic product in 1987 (28.5%) is almost the same as it was in 1970. As
a result of the import substitution/capital-intensive development strategy, labor has not been able to shift to industry, and what labor has moved out of
agriculture has more often than not ended in the service sector (David 1983).

On a more general level, agriculture has suffered vis-à-vis manufacturing owing to the overvalued exchange rate and government investment and
price policies which have deliberately kept the price of rice low (APST 1986).
Between 1972 and 1980, the terms of trade between the rice price and the non-
food price index declined from 1.0 to 0.59 (Hill and Jayasuriya 1984). Thus,
what growth there has been in the agricultural sector came not as the result of
but in spite of government policies (David 1982; Rocamora 1979). In addition,
landlessness and near landlessness in rural areas have been reported to be
more than 75% (Rosenberg and Rosenberg 1980) and landlessness among agricultural farm laborers is almost 50% (APST 1986; Porter and Ganapin 1988).
Finally, land reform has been ineffective in actually transferring land to the
peasants because of bureaucratic delay and widespread erosion of the spirit of
the law (Carroll 1983; ILO 1974; Kerkvliet 1974; Tiongson et al. 1986; Wurfel
1983). It is interesting in this regard that land distribution has yet to occur on
the property of the family of President Aquino.

Table 19 presents data on the number and area of farms. Note that farm
area as reported in the various censuses is less than harvested area (primarily
owing to double and triple cropping of rice and corn) and more than cultivated land area (since not all farmland is actually under cultivation at any
one time).

While the data presented in table 19 are useful, they do have several
major limitations. First, the agricultural censuses do not include information
on ownership. Second, it is impossible to tell whether farms are being subdivided, for example, as a result of inheritance or to avoid land reform, and it is
equally impossible to tell whether farms are being enlarged, for example, by
wealthier farmers or corporations buying out smaller farms. Most likely both

Table 19.
Farm area, 1948–80

Year	Number of farms	Increase in farms from previous year	Farm area (km^2)	Increase in farm area from previous year (km^2)	Average farm size (ha)
1948	1,638,624	—	57,266	—	3.49
1960	2,166,216	527,592	77,725	20,459	3.59
1970	2,354,469	188,253	84,937	7,212	3.61
1980	3,420,323	1,065,854	97,252	12,315	2.84

Sources: Agricultural censuses, 1948, 1960, 1970, 1980.

processes are occurring to a certain extent, but the data do not permit a detailed examination of these trends (Rosenberg and Rosenberg 1980).

The important issue is whether or not concentration of agricultural lands is taking place. Rosenberg and Rosenberg (1980) feel that it is, as a result of the increased demand for agricultural goods and the new agricultural technologies. In addition, with the shift to commercial agriculture, plantations are expanding and forcing poorer farmers off the land. This would be the case particularly in Mindanao, which is home to large banana and pineapple plantations (APST 1986; Costello 1984; Tiongzon et al. 1986; Van Oosterhout 1983). Belsky and Siebert (1985) claim that in Leyte commercialization and concentration of agriculture in lowland areas are decreasing the amount of land available for poor farmers and forcing them to initiate farming in upland areas. Luning (1981) states that the expansion of sugar land in the western Visayas from 1960 to 1975 occurred primarily at the expense of upland rice and corn lands, and Krinks (1974) details increasing land concentration in a frontier region in southern Mindanao. In short, there is evidence to support the proposition that with the increasing spread of commercial agriculture in the lowlands, the ownership of agricultural lands is becoming more concentrated in the hands of wealthier farmers and corporations. At the same time, small farms are becoming smaller (Luning, 1981), most likely as the result of subdivision through inheritance. The end result has been increasing landlessness for the rural poor (Cruz and Zosa-Feranil 1988).

When interpreting table 19, it should be kept in mind that according to the Bureau of Soils (1977), arable land in the Philippines is 84,362 km^2, with arable land defined as land that can be sustainably farmed on an annual basis with little or minimal investment in land conservation (A, B, C, and D lands in the Bureau of Soils categorization). Farmland in 1970 was slightly more than total arable land. In addition, since "urban areas and others" now take up approximately 11,000 km^2 (APST 1986) and much of this land is most likely arable land (Bruce 1987; Gwyer 1978), this means that at least 20,000 km^2 of farmland in 1980 was not on arable land. If we assume that farmers prefer arable to nonarable land for agriculture, then most of the increase in farm area since 1960 has been on nonarable land, as defined by the Bureau of Soils.

During the 1960–70 period the number of farms increased by only 188,253 (8.7%), farm area increased by 7,212 km^2 (9.3%), and average farm size remained virtually the same. A tentative conclusion would be that the "land frontier" had been reached during this period, and it is to this issue that we turn our attention next.

The land frontier appears to be that areal limit beyond which agriculture cannot be safely practiced (APST 1986; Costello 1984; Hayami et al. 1976; Huke 1963; ILO 1974; Rosenberg and Rosenberg 1980; Ulack 1977; World Bank 1976). Although this is not a rigorous definition, none of the sources cited above defines the term in a precise manner with the exception of Hayami et al. (1976), and their usage is based primarily on economics. In the literature just cited, the land frontier would be at the margins of arable land. Unfortunately, arable land itself is a vague concept.

For instance, to what extent does the determination of the limits of arable land depend upon cultural practices or investments in agricultural land? Does it refer to annual crops or does it encompass perennials also? Is slope the major determining factor in whether land is arable or not? Most important, does the concept of arable land or a land frontier make any sense when applied to subsistence agriculture? In the context of the Philippines, the questions above remain unanswered; however, the fact that numerous authors have reported serious and widespread land degradation in the uplands means that the term may have some value when applied to land-use practices which result in substantial off-site effects. At the same time, land degradation does not necessarily mean that the land frontier has been reached; it could simply mean that land-preserving investments, e.g., terracing, have not been made. This then raises the question of why land-preserving investments are not being made, an issue that is central to the analysis of Blaikie (1985) and Blaikie and Brookfield (1987).

Kikuchi and Hayami (1978) argue that the Philippines shifted from extensive to intensive cultivation during the 1950–69 period. As the land/labor ratio declined, the growth rate of cultivated land slowed and the Philippine government was forced to invest in irrigation. In short, there was a shift from "external land augmentation" to "internal land augmentation." Huke (1963), using a maximum slope of 20% for cultivable land, claimed that the land frontier in Mindanao would be reached in the early 1970s, and this sentiment was shared by Wernstedt and Simkins (1965). Hooley and Ruttan (1969) proclaimed the closing of the land frontier in the 1960s, as did Cant (1979). The ILO (1974, 83) predicted the closing of the land frontier and concluded: "The mission doubts that land expansion can ever again make the sort of contribution it made before 1960." This conclusion was shared by the World Bank (1976). Reynolds (1985, 184) stated that by 1960, "the possibilities of acreage expansion were largely exhausted."

In short, there was widespread agreement that by the late 1960s or early 1970s, the Philippines had reached some sort of land frontier and that future growth of agricultural output would have to come from increases in productivity rather than from increases in the area of production. In fact, what happened is that agricultural output increased, productivity increased, and the area under cultivation also increased considerably. As table 19 indicates, in the period from 1970 to 1980, the number of farms increased by 1,065,854 (45.3%) and farm area by 12,315 km^2 (14.5%). As a result, average farm size declined from 3.61 ha to 2.84 ha (21.3%). The increase in farm area between 1970 and 1980 (12,315 km^2) was 70% greater than from 1960 to 1970 (7,212 km^2), and Feder (1983) attributes this to the rapid expansion of commercial crops. The continued decline of forest area in the 1980s surely indicates that the area of farmland is continuing to increase. In short, this discussion raises serious questions as to how useful the concept of the land frontier is and introduces the more relevant questions of how much land is suitable for agriculture and what type of agriculture it is suitable for (Gwyer 1978).

NEDA (1981) points out πthat lands classified A, B, C, and D by the Bureau of Soils in 1976 were 83,500 km^2, and farmland was 84,890 km^2. However, they suggest that lands suitable for agriculture total 145,850 km^2 and this includes lands appropriate for grazing, fishponds, and annual and perennial crops. Luna (1966) is in general agreement with this conclusion and estimates that 150,000 km^2 of the Philippines is arable. On the other hand, Gwyer (1977) calculates that 238,871 km^2, or 80% of total land area is suitable for agriculture, with cultivable land equal to 83,300 km^2, intermediate land suitable for pasture or grazing equal to 149,840 km^2, and wetlands appropriate for fishponds equal to 5,570 km^2. In the uplands alone, Cruz and Feranil (1988), using a cutoff of 30% slope or less, have determined that approximately 84,000 km^2 could be suitable for agriculture.

The work of Gwyer (1977, 1978), NEDA (1981), and Cruz and Feranil (1988) make it clear that the notion of a land frontier based on arable or easily and safely cultivated land may not be appropriate for Philippine conditions. If all arable land is presently being used and if the supply of arable land will decrease with the expansion of human settlements in the lowlands, then this necessarily implies that environmentally sound agricultural expansion will have to rely on improved pastureland management, perennial crops, or large investments in land improvement. Whether or not this is happening on a large scale is not known, but the fact that in 1982, 2.5 million ha of cropland were on slopes greater than 18% (APST 1986) means that it has already become an issue of national importance. That perennial tree farming in the Philippines is "relatively neglected" (NEDA 1981) and the numerous observations of land degradation in the uplands may indicate that cultivation of annuals is widespread. The failure to plant long-term crops may simply reflect the lack of security of tenure (Krinks 1970) or the fact that annual food crops are more important to subsistence farmers (Belsky 1989).

In a general sense, upland areas are defined as those lands with slopes of 18% or greater and they comprise approximately 55% of the Philippine land area. They are important for three reasons. First, since most public forests are on lands with slopes of 18% or greater, they contain the majority of the remaining Philippine forests. Second, they have been the destination of numerous migrants in the postwar period. Third, land degradation in the uplands is severe and widespread. However, a great deal of uncertainty exists regarding the uplands, particularly in relation to the extent and nature of agriculture.

Cruz et al. (1986) estimated that 14.4 million people lived in the uplands in 1980 and 77% of these were on lands officially classified as public forestlands. This figure should be contrasted with the 1,327,359 occupants listed by the DENR as living on forestlands in 1986 (*PFS* 1986). During the 1948–80 period, upland population grew between 2.5 and 2.8% a year. While this is less than the national population growth rate, upland areas have higher mortality and lower birthrates than lowland areas (Cruz 1990). According to Cruz et al. (1986, 40), "Migration accounts for the bulk of the population growth." Lynch and Talbott (1988) argue that of the roughly 18 million people living in the uplands in 1988, 6 million were there before 1945, 2 million migrated between 1945 and 1948, and 10 million have migrated since 1948. In addition, high rates of migration to the uplands have continued in the 1980s (Cruz 1990; World Bank 1989a). Interestingly, the highest rates of population growth in the uplands are in those municipalities which have logging concessions (Cruz and Zosa-Feranil 1988).

Almost all observers agree that this migration is occurring because of the lack of opportunities in the lowlands. In short, poor people are being forced to the uplands because they have no choice; according to Cruz (1988), 70% of all upland migrants are landless lowlanders. As Westoby (1981) points out, these poor farmers should be properly referred to as "shifted cultivators."

The SSC (1988) study of natural vegetation evaluated extensive and intensive land use and forest cover. It was the first and only study to cover all types of land uses. Extensive land use was defined as areas with more than 10% but less than 70% cultivated land and can be considered to be synonymous with farming on steep slopes (World Bank, 1989a). Intensive land use was defined as areas that were more than 70% cultivated.

To the reservations expressed in chapter 3 about the SPOT interpretation of forest cover must be added two serious qualifications regarding the ability of SPOT (or LANDSAT) imagery to evaluate cultivated land accurately. First, most annual crops (with the exception of irrigated rice) are grown in the wet season and therefore they may not be obvious on dry-season images. Second, on a dry-season image, the distinction between cultivated land and grassland may be difficult to make (World Bank 1989a).

Given this caveat, the SSC results indicate that 41% of the Philippines or 119,576 km^2 was under extensive cultivation and 34% or 99,340 km^2 was under intensive cultivation in 1987. If fishponds are subtracted from land under intensive cultivation, then intensively cultivated lands equal 97,287 km^2 in 1987. This is almost exactly the amount of farmland presented in table 18 for 1980.

The World Bank (1989a, annex 1, table 4), on the basis of the SSC results, calculated total area cultivated by assuming percentage of agricultural use in each of the land-use categories as follows: 100% in cultivated areas in forests, 30% in grasslands, 40% in mixed cultivation/brushland/grassland, 85% in arable land, 85% in mixed cropland and plantations, and 100% in fishponds. The result is that cultivated land equals 11.3 million ha or 38% of total land area. By category, upland agriculture, plantation crops, annual crops, and fishponds take up 13, 8, 16, and 1% respectively of total land area. Cultivated area in the uplands is approximately 3.9 million ha. In general agreement with the notion that cultivated land in 1987 may be 11.3 million ha are data from the Bureau of Agricultural Economics (1987) which show 11.9 million ha of cropland in 1985.

The World Bank (1989a) admits that these results are "plausible" but contends that the margin of error is "certainly wide." Since one of the major difficulties in discussing land-use changes in the Philippines, given the data, is that it is fairly easy to arrive at "plausible" estimates with wide margins of error, it is not clear how the above results are to be interpreted. More important, it is not at all clear how the figures are to provide a guide for remedial action on the part of the government. In any case, it would appear that agriculture in the uplands is widespread. Myers (1988a) states that total farmland covers 13 million ha, of which 5.4 million are in the lowlands and 7.6 million are in the uplands. Unfortunately, no source is given and it was not possible to assess the accuracy of these claims.

It is interesting to compare the World Bank figure of 11.3 million ha of cultivated land in 1987 with the census data in table 19 which indicated 9.7 million ha of farmland in 1980. If both data sets are correct, then cultivated land increased by more than 1.6 million ha between 1980 and 1987. This is equal to 229,000 ha per year, a figure I consider to be reasonable. By way of contrast, I calculated (table 7) that the average annual rate of deforestation between 1980 and 1987 was 157,000 ha per year. If correct, this necessarily means that approximately 70,000 ha of grassland/brushland was being converted to agriculture each year. This could be a further indication of the increasing pressure on upland land resources.

One of the most serious gaps in our understanding of what is happening in the uplands to the agriculture/forest interaction has to do with shifting

Table 20.
Area deforested annually by *kaingineros*

Step	Output
1 Average size of *kainginero* family	5.2 persons
2 Length of *kaingin* cycle	12 years
3 Number of *kainginero* families	1,019,230
4 Area of forestland per family	3.9 ha
5 Area of land cleared per family per year (3.9 ha/12 years)	0.325 ha
6 Total area deforested per year (1,019,230 x 0.325)	331,250 ha

Source: Bee (1987, table 11).

cultivation. In the Philippines, the Tagalog term *kaingin* is generally assumed to be equivalent to shifting cultivation, but as Olofson (1983) has pointed out, this one-to-one correspondence is not correct. Agriculture in the uplands can consist of traditional shifting cultivation (long fallow periods), nontraditional/migrant shifting cultivation (short fallow periods), permanent/intensive agriculture, backyard gardens, grazing, or any combination of these. Unfortunately, there is no reliable information on how extensive these forms of agriculture are. In addition, there are no reliable data on where shifting cultivation is taking place, whether it be in grasslands, brushlands, or secondary or primary forest. Finally, there are no data at the national or provincial level on how often farmers shift their plots. If we accept as a working definition that shifting cultivation is a system of agriculture characterized by a short period of cultivation followed by a long fallow period (Lanly 1985), then it will be the contention of this study that shifting cultivation is not as widespread as is commonly assumed.

Bee (1987) has attempted to determine the area deforested annually by *kaingineros* in the Philippines, and table 20 is an abbreviated summary of his calculations.

While table 20 and the assumptions and calculations that went into Bee's work could be discussed in considerable detail, it is not my intent to judge unfairly what is admittedly a very rough estimate (although see Eder 1989 for a critical review of this work). However, given the lack of data which confronts all researchers in this field and the assumptions needed to arrive at a final figure of deforestation per year, two crucial steps in table 20 require comment. We will demonstrate how precarious Bee's final figure is by changing the numbers in steps (2) and (3).

Bee assumes an average *kaingin* cycle of twelve years. While this may be true for traditional shifting cultivation, it certainly is not for migrant shifting cultivation (Cruz et al. 1986; Cruz and Zosa-Feranil 1988; FAO/UNEP 1981; Fujisaka 1986). In fact, if the average *kaingin* cycle were twelve years, then it is possible that many of the negative off-site effects of upland cultivation would not be as serious as they are. For the sake of argument, assume an average *kaingin* cycle of six years. Bee's work was conducted in 1981 and at that time it may have been appropriate to assume 1 million *kaingin* families. However, the recent work of Cruz et al. (1986) indicates approximately 2.5 million families in the uplands in 1980. For the sake of argument, assume 2 million families occupying the uplands.

If we retain Bee's method of calculation but use an average *kaingin* cycle of six years and 2 million *kaingin* families, the result is 1.3 million ha deforested per annum. Such a high rate is simply impossible (if true, all forests would have been destroyed years ago), and demonstrates how sensitive deforestation figures derived in this manner are to the initial assumptions and data. In general, Bee's work raises very serious questions as to what is happening in the uplands with regard to shifting cultivation and deforestation.

To my mind, it is very difficult to reconcile the upland population figures of Cruz et al. (1986) with the rates of deforestation calculated in chapter 3 and the commonly accepted notion that almost everyone in the uplands is practicing shifting cultivation. Since the data on forest cover and population are taken as being reasonably accurate, this means that there are logical grounds for doubting that shifting cultivation is as widespread as assumed. If this is the case, it necessarily means either that not all uplanders are engaging in agriculture or that, if they are, it is not necessarily shifting agriculture. Since data do not exist on the occupation of uplanders, it is impossible to determine with any degree of certainty how many people are practicing agriculture. However, there is some evidence to indicate that forms of agriculture other than shifting cultivation exist in the uplands.

Vandermeer (1963), in a study of Cebu province, which is now entirely deforested, points out that what had originally been a system of migratory corn cultivation is now basically sedentary corn cultivation. The main impetus for the change was the increasing population density. In a case study of a pioneer community in Laguna province near Manila, Fujisaka (1986) indicates that an original system of shifting cultivation is now evolving to a more permanent, mixed system of agriculture. Cornista et al. (1986) point out that some upland farmers are starting to plant trees and perennial crops. In northern Luzon, near the major city of Baguio, intensive truck farming of temperate crops is widespread and entails the heavy use of modern agricultural inputs such as pesticides and fertilizers (Russell 1989). Eder (1977) demonstrates that agricultural intensification is taking place on Palawan, and Olofson (1984) has reported "stabilization of shifting cultivation" among the Ikalahan of northern Luzon. In addition, Olofson (1980) reports little evidence of shifting cultivation in the province of Laguna. Beets et al. (1986), in a study of development projects for NOVIB-Philippines (a Dutch nongovernmental organization), note that 30–70% of all farms in the projects studied were "more or less permanently established farms." Lastly, Cruz (1984) and Cruz and Zosa-Feranil (1988) present several examples where commercialization of agricul-

Table 21.
Deforestation and its relation to increases in population and farmland, 1948–80

Years	(1) Increase in farm area (km^2)	(2) Increase in population (millions)	(3) Loss of forest cover (km^2)	(4) 3/2*	(5) 3/1**
1948–60	20,459	7.831	– 25,073	.0032	1.2
1960–70	7,212	9.596	– 22,465	.0023	3.1
1970–80	12,315	11.416	– 21,032	.0018	1.7

Sources: Tables 7, 17, and 19.

Notes: The area of forest cover in 1948 was assumed to be 150,000 km^2. Forest cover in 1960 was determined by the straight line method using the NEC data for 1957 and the P-GFI data for 1969. Forest cover for 1970 was determined by the straight line method using P-GFI data for 1969 and 1980.

* Column 4 is the area deforested divided by the population increase for the same period. It represents hectares of forest lost per one person increase in population.

** Column 5 is the ratio of the area deforested to the increase in farm area.

ture in the uplands is leading to a more intensive farming system. In short, there is evidence to indicate that more permanent forms of agriculture are starting to emerge in the uplands. Table 21 shows the relation of deforestation to population and agriculture in a macroperspective. Column 4 shows a steady decrease in the area of deforestation for each 1 million increase in the population. Since population is increasing absolutely, this does not necessarily mean that deforestation is slowing down. Column 5 indicates that deforestation in the postwar period until 1980 was greater than the expansion of farmland. The 1960–70 period stands out in particular in this regard.

Table 21 covers only 1948 to 1980. It indicates that deforestation was 23% to 300% greater than the spread of agriculture in each of the three time periods. On the other hand, the comparison between deforestation and the spread of agriculture in the 1980s indicated that the latter may now be greater than the former by approximately 70,000 ha per year. If so, it may be the case that the spread of agriculture is now occurring in grasslands. While this conclusion is tentative, it may provide an interesting insight into the forest/agriculture interaction.

"Forest/agriculture interaction" is too simplistic a term to capture adequately the process of deforestation and postdeforestation land use. A more accurate term would be "forest/agriculture/grassland interaction," with grassland comprising open land, brushland, shrub land, wasteland, and abandoned land. At this point, we simply note that Grainger's model of land use (1987), which assumes a one-to-one correspondence between deforestation and the expansion of agriculture, would require modification if this is correct.

Lastly, if it is true that after 1980 the expansion of agriculture became greater than deforestation, then this event could be of significance in and of itself. It could signal a true intensification of pressure on land and forest resources. As forest area declines and human settlements increase, a point is reached where forest conversion can no longer meet the need for new agricultural land (as in the pre-1980 period) and former grassland is then converted to agriculture. While only a tentative conclusion, this line of reasoning highlights the validity of the notion of a forest/agriculture/grassland interaction.

Summary

Overall, the picture presented in chapter 4 is of a postwar development strategy which has failed to provide jobs in the cities and has left the majority of Filipinos impoverished. While there are serious gaps in the data, it seems safe to conclude that land under agriculture is continuing to expand, and increasingly, if not entirely, it is occurring in the ecologically fragile uplands; in some areas of the uplands, agricultural intensification is occurring and migration to upland areas can be expected to continue as concentration of landholdings proceeds in the lowlands. Major areas of uncertainty are the precise nature of the forest/agriculture/grassland interaction and the relative extents of shifting and permanent agriculture. In addition, there is virtually no reliable information on how extensive commercial agriculture is in the uplands or on types of land-preserving investments that agriculturists may be making.

In short, the major process that has occurred in the past forty-five years in the Philippines has been the impoverishment of its people. Much of the expansion of agriculture and migration has been the result of desperately poor people trying to find a place where they can at least feed their families. Owen (1983, 197) stated the reasons behind these activities when he noted: "It would be hard to prove that the material welfare of the average Filipino is significantly higher today than it was in the nineteenth century."

5
The Deforestation Process in the Philippines

Understanding Tropical Deforestation

A clear-cut summary of the causes of tropical deforestation is not possible. The major problem is that deforestation is the end result of a process which occurs at many levels and there are numerous connections between and within the various levels. The discussion in chapter 2 indicated that the four main agents of forest destruction, on a worldwide basis, are agriculturists, ranchers, loggers, and fuelwood collectors. These four groups are the ones who actually cut down the trees. It is important to note that all four agents of deforestation require access to the forest and in almost all cases this is provided by road networks. In short, it appears logical to expect road networks to be part of the deforestation process.

In a general sense, the motive of the agents of deforestation can be either monetary gain or subsistence. Much of the expansion of agriculture into previously forested areas and much fuelwood gathering is by poor members of society whose primary concern is subsistence (see Belsky 1989 for a discussion of the importance of access to a secure source of staple food for poor upland farmers). On the other hand, logging, some agriculture, and some charcoal making are primarily for commercial purposes. In short, different groups of people cut down the forests for different reasons. Loggers and fuelwood collectors are interested in the forest per se, ranchers in the land that forests occupy, and agriculturists in both, since conversion of the surface biomass to ash is a vital part of the agricultural production process. Shifting cultivation occurs primarily in forests because burning the forest supplies fertilizer to the soil. In other words, shifting cultivation can be viewed as a system of agriculture which moves to where there is a free supply of fertilizer; a feature which makes it very attractive to those who are poor.

More fundamentally, for deforestation to occur, forests must exist and they must be accessible. The characteristics of the forest of most importance would be (1) extent and geographic location, (2) structure (primary or secondary forest), and (3) economic value (species composition and presence of minor forest products such as rattan or bamboo). Accessibility to the forests in turn will be a function of (1) infrastructure, (2) rules governing access, and (3) penalties for not following the rules governing access. The last point is particularly important because if there are no negative sanctions for not following the rules regarding access to the forest, then there are no rules at all.

The encroachment of subsistence activities (agriculture and fuelwood gathering) onto forestlands is, to a large extent, the result of poverty and lack of alternative employment opportunities. While evidence on the relative extent of traditional shifting cultivation versus shifting cultivation practiced by recent migrants is almost completely lacking, I am assuming (at least for the Philippines) that the latter is by far more extensive (FAO/UNEP 1981). In turn, accessibility of the poor to forests is increased through the construction of infrastructure (including logging roads) and through poor or nonexistent government regulation of access to forests. Commercial use of forests (logging, ranching, charcoal making, commercial agriculture), on the other hand, is primarily by better-off members of society (or foreigners) and is sanctioned by the government through concessions or, in the case of ranching in the Brazilian Amazon, through subsidies. Overall, use of the forest by the poor is almost always illegal and commercial use of the forest is almost always legal. In the case of the Philippines, Makil (1984) notes that use of the forest for profit has always taken precedence over use of the forest for subsistence. The main issue is, "Who has the right to use the forest resource?" (Byron and Waugh 1988) and in the Philippines it has almost invariably been logging concessionaires.

The recognition of differential access to the forest by poor and rich groups and the existence of large numbers of people who have no alternative but to practice subsistence agriculture raises questions about the process of development itself, elite control of government, and the use of public office for private ends. This is a step removed from the actual cutting of trees, but it is the context in which the agents of deforestation operate.

Table 22 presents six major users of the tropical forest, their activities, and the results. "Logging" represents selective logging conducted in an environmentally sound manner. Table 22 is a broad generalization of processes occurring worldwide and, as a result, counterexamples can easily be found.

"Equilibrium?" in table 22 refers to whether or not the process of deforestation stops or must be repeated. If the process must be repeated, the time interval may be as short as a month (commercial charcoal making), a year

Table 22.
Activities leading to tropical deforestation

Activity	Nature of forest		Result	Equilibrium?
	Secondary	Primary		
Permanent agriculture	X	X	Deforestation	Yes
Shifting cultivation	X	X	Deforestation by migrants	No
Ranching	X	X	Deforestation	No
Logging		X	Deforestation and creation of secondary forest	No
Fuelwood gathering	X	X	Deforestation in dry tropics; Degradation in wet tropics	Yes; No
Charcoal making	X	X	Deforestation	No

(some types of shifting cultivation), or five years (ranching). The only activity which leads unequivocally to an equilibrium position after deforestation is permanent agriculture. However, even this must be tempered by the observation that large-scale permanent agriculture can also displace poor farmers previously on the land and these displaced peoples may then be forced to cut down forests in order to practice agriculture. Fuelwood gathering, in the wet tropics, does not appear to result in deforestation to any significant extent.

The relationship between activity and result is a bit more complicated than indicated in table 22. Logging can directly result in deforestation if clear-cutting is practiced; however, it appears that more often than not it leads to a degraded or secondary forest. The secondary forest created by logging can then be used by the other five activities, and this almost invariably results in deforestation.

At one level removed from the actual agents of deforestation, three major factors are emphasized by most authors: increasing populations, widespread poverty and income inequality, and the open-access nature of the forest resource. These considerations provide the socioeconomic framework in which the agents of deforestation operate. An additional factor, particularly in the Philippines, has to do with elite control of government and widespread corruption in the forestry sector.

The Deforestation Process in the Philippines

Figure 2 is a highly simplified box-and-arrow diagram of the major forces leading to deforestation in the Philippines. While I recognize that some deforestation has been caused by other factors like charcoal making, it is postulated that the two most important activities leading to deforestation are logging (legal and illegal) and the expansion of agriculture. As Gillis (1988) points out, both of these factors must be considered together, along with rural poverty and the open-access nature of the forest resource, which is primarily the result of government policy. In short, the deforestation process in the postwar Philippines can be characterized by two major steps: the conversion

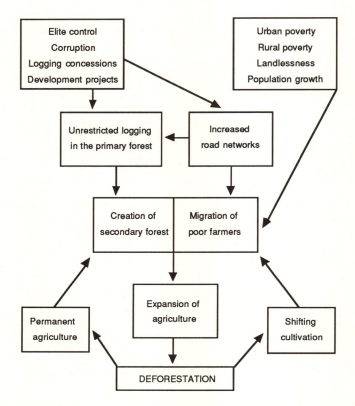

Fig. 2. *Overview of deforestation in the Philippines*

of primary to secondary forest by logging activity and the removal of secondary forest cover by the expansion of agriculture.

Figure 2 assumes that logging does not directly result in deforestation; in other words, the assumption is that selective logging, properly practiced, converts a primary forest into a degraded secondary forest. This is an unrealistic assumption to make for two reasons. First, it is likely that clear-cutting has been practiced in certain areas, although Gillis (1988) claims that this has been "relatively rare" in Southeast Asia. On the other hand, Walker (1987) claims that clear-cutting is widespread in Indonesia. Unfortunately, data on the relative extent of clear-cutting versus selective logging in the Philippines do not exist. Second, given the extensive road networks and yarding areas needed for capital-intensive logging and extensive damage reported for some logging operations (Blanche 1975; Burgess 1971, 1973; Egerton 1953; Gillis 1988; Philippine Council for Agriculture and Resources Research and Development 1982; World Bank 1989a), some selective logging has directly resulted in deforestation. In addition to these considerations, it seems reasonable to assume that illegal commercial logging has been even more destructive than legal commercial logging. In short, the assumption in figure 2 that logging itself does not lead directly to deforestation is a simplifying step and made for purposes of presentation only. It is important to note that "logging" refers to legal and illegal logging.

The relationship between logging and the primary and secondary forests is a dynamic one. As logging converts the primary forest to a secondary forest, loggers move on to new primary forest. Implicit in this schema are the notions that secondary forests do not return to a primary state and that loggers do not return to log the degraded forest they created in the first place. While these generalizations are not always true (the author is aware of several firms in the Philippines who have returned for a second cut), as overall statements they fit the Philippines well. Also implicit in this framework is the notion that concessionaires have not, in general, engaged in protection of the secondary forest, enrichment planting, or reforestation. While this is a strong statement, forest protection has been minimal in the Philippines (FAO/UNEP 1981).

In short, I postulate that the major cause of the decline of the primary forest has been logging. As the World Bank (1989a, 7) points out: "It is mainly due to logging (licensed or illegal) that the old-growth dipterocarp forests, the most valuable commercially, have shrunk from 10 million hectares in the 1950s to only one million today."

The assumption that the expansion of agriculture takes place primarily in secondary forests seems warranted for three reasons: first, on logical grounds, it is much easier to clear secondary than primary forests (Byron and Waugh 1988) (in an economic sense, as Southgate and Pearce [1988] point out, logging lowers the land-clearing costs of settlers); second, degraded forests are more likely to be penetrated by roads than primary forests and roads greatly facilitate the expansion of agriculture; third, numerous observers have commented upon the spread of agriculture in secondary forests and logged-over areas and, as I pointed out above, most of the destruction of the primary forests has been the result of logging (ADB 1976; Segura-de los Angeles 1985; Edgerton 1983; FAO/UNEP 1981: Hackenberg and Hackenberg 1971; Vandermeer and Agaloos 1962; van Oosterhout 1983). In short, I accept the observation of Hicks and McNicoll (1971) that almost all deforested land has been converted to agriculture and that most of the expansion of agriculture has occurred in former selectively logged forest areas as a generally accurate statement regarding changing land use in the Philippines.

The relationships among the expansion of agriculture, the creation of secondary forest, and deforestation are also dynamic. Deforestation is the result of agriculture encroaching on secondary forest after logging has occurred; however, the time lag between logging and the spread of agriculture is not known. If logging is followed by shifting agriculture, then the process will have to be repeated in one to three years; if logging is followed by permanent agriculture, then the individual farmer will not change to a new location. The importance this study places on the relative extent of shifting versus permanent agriculture on recently deforested lands stems from the fact that farmers practicing short-term agriculture must necessarily move to another geographic location. As pointed out in chapter 4, there is some evidence to indicate that intensification in the uplands is occurring. However, detailed information on this point is still lacking.

Preceding logging and the expansion of agriculture is the construction of roads (Hackenberg and Hackenberg 1971). These roads are primarily the result of development considerations by provincial and national governments or loggers who have concessions. As such, roads can vary from little more than dirt tracks to paved highways. While roads can facilitate the spread of agriculture by opening up new areas—for example, this seems to have occurred in parts of Mindanao in the 1950s and early 1960s (Simkins and Wernstedt 1971; Vandermeer and Agaloos 1962; Wernstedt and Simkins 1965)— roads followed by logging make agriculture even more possible by converting the primary forest to a degraded forest. In addition, logging provides jobs and thus directly leads to an increase in population where it is occurring. The relationship of new roads to deforestation in Thailand has been clearly drawn by Thung (1972) and in Brazil by Fearnside (1986).

A caveat to this discussion of roads must be mentioned: given the island nature of the Philippines, it is conceivable that logging near the coastline occurred with only minimal construction of roads. While this may have happened in some areas, even those logging concessions located directly on the coast have large road networks (personal observation of author; Egerton 1953) and as Luna (1966) points out, the inland forests of Luzon and Mindanao require "extensive land transportation facilities." As such, large-scale logging without road networks is not considered to be a widespread occurrence.

Lastly, a caution regarding the relationship between roads and deforestation must be introduced. The authorities cited above have all claimed that roads precede the expansion of agriculture. I would qualify this statement to a certain extent. First, it may be the case that this is true primarily for large road networks which open up previously low-populated areas such as the Brazilian Amazon. If so, then smaller or secondary roads may not be as important for explaining deforestation. Second, there are cases in the Philippines where agriculture has preceded roads. In short, the road/deforestation relationship may not be as clear-cut as has been presented above.

The expansion of agriculture onto forested lands is driven by three major forces: the increase in population, widespread poverty, and the political economy of lowland agriculture. An additional possibility is that deforestation is the result of the expansion of commercial agriculture by people who are not poor and are producing for the market. Data on this point are lacking, however, and the evidence presented in chapter 4 that sedentary/commercial agriculture may be occurring in certain upland areas cannot be used in a quantitative analysis. The assumption of this study is that most of the deforestation committed by agriculturists is done by subsistence farmers (traditional and migrant). As an example, Segura-de los Angeles (1985), in a case study of an upland agroforestry project in Luzon, notes that 88% of all those surveyed consumed all of the rice they produced ˙and did not have a surplus of marketable vegetable production. Hackenberg and Hackenberg (1971), in a study on development in Davao, note that while farmers were growing some commercial crops, the primary crops were rice and corn.

The increase in poverty (relative and absolute) in the Philippines in the past forty years is evidence of the failure of development to improve living standards for the bottom 50–75% of the population (Gorra 1986). As a USAID study (1980) pointed out, at least 56% of all Filipino families had an income below that required for a "minimum nutritionally adequate diet." Further, they note, "The overriding goal of the low income household is to produce or earn enough to eat. A food income in kind provides a certain security" (p. 5). In short, poor people are forced to engage in subsistence agriculture because that is the only option available to them (Gwyer 1978).

The granting of concessions, on the other hand, has occurred for two reasons: first, from the legitimate desire of the Philippine government to foster development and, second, as political favors to either Philippine elites or multinational corporations (primarily American in the 1950s and 1960s). I would be hard-pressed to argue that the postwar Philippine government has ever really been concerned with development in the forestry sector; rather, it would appear that forests were viewed as an asset whose benefits should flow mainly to politicians and the well-connected (Ofreno 1980; Palmier 1989). As Hackenberg and Hackenberg (1971, 8) point out in their study of Davao City, Mindanao, "The basis of wealth is lumber and the profits are instantaneous for those with political connections to secure a concession." In fact, the distinction between politicians and loggers is difficult to make, since loggers contribute heavily to political campaigns and many politicians control logging concessions (*Economist* 1989).

Summary

In an overall sense, deforestation in the postwar Philippines is the result of two major processes: the conversion of primary to secondary forest through logging, and the removal of secondary forest cover by the expansion of agriculture. Both processes require the construction of roads; in addition, the spread of agriculture requires the movement of subsistence farmers to degraded secondary forestlands. The virtually unrestricted access to the dipterocarp forests by concessionaires has also helped to create and sustain the conditions which led to impoverishment for the majority of Filipinos: elite control of government.

6
Data Sources

Introduction

The statistical analysis to be conducted in chapter 7 will comprise two parts: first, a cross-sectional examination using multiple regression techniques will be conducted for 1957, 1970, and 1980; second, panel and path analysis will be used to examine changes in forest cover from 1970 to 1980. In the cross-sectional analysis, the dependent variable will be the absolute amount of forest cover in each province; whereas, in the panel analysis, the dependent variable will be the amount of deforestation in each province from 1970 to 1980. In both cases, the independent variables will be the socioeconomic and physical factors which are the subject of this chapter.

General Data Problems

Provincial-level data on socioeconomic and physical indicators in the Philippines are difficult to come by. For all practical purposes, longitudinal data at the provincial level do not exist, with the exception of several variables explained below. Thus, I am forced to concentrate on three years only, 1957, 1970, and 1980. The variables reviewed involve forest cover, demography, agriculture, infrastructure, forestry, and geophysical attributes.

One of the primary problems encountered in obtaining a consistent set of data was changing provincial and regional boundaries. The Philippines currently has seventy-three provinces and thirteen regions (including the National Capital Region as a separate region), but in 1970 there were sixty-six provinces and ten regions and in 1957 there were fifty-four provinces and nine regions. In many cases, the relevant data have been adjusted retroactively by the appropriate government agency to reflect boundary changes (for instance, population statistics), but in some cases data for 1970 and 1960 may be for a province which no longer exists and, therefore, are not compatible with present-day provincial boundaries. Fuchs and Luna (1972) encountered a similar problem when comparing 1939 and 1960 census data. Fortunately, in most cases, earlier provinces were divided into smaller provinces. For example, Agusan del Norte and Agusan del Sur in 1970 are equal in area to Agusan in 1960. In this case, any data from 1960 would be allocated between Agusan del Norte and Agusan del Sur on a proportional basis according to either area or population in 1970. Variables proportioned on the basis of area were arable land, timber production, number and size of concessions, and ACC. Variables proportioned on the basis of population were agricultural area, number of farms, kilometers of roads, and urban population.

Overall statements regarding the quality of data are difficult to make. Population census data should have a high degree of reliability; on the other hand, data on the forestry sector and road networks have serious shortcomings, discussed in their respective sections. A potentially more serious problem is simply the lack of data. This means either that the data were never collected or were collected and are now missing. This issue will be discussed below for individual variables. Two examples should suffice to demonstrate some of the difficulties involved in working with Philippine data.

In 1983, the Data Assessment and Review Team of the Ministry of Agriculture published the *Regional Statistical and Agricultural Profile* for all thirteen regions of the Philippines. In volume 4 (*Southern Tagalog*), table 14.01 (p. 14–1) indicates that total highway kilometerage was equal to 18,352. However, data supplied to me by the Department of Public Highways in 1988 show that the figure was actually 11,140. The difference is substantial and, as explained below, even given the shortcomings of the Department of Public Highway's data, I had no reasonable alternative to using their figures.

In 1985, the Bureau of Agricultural Economics published a profile of region 5 (Bicol), and their data (table 136, p. 187) indicate that the service areas of communal and pump irrigation systems were 58,881 and 28,040 ha respectively. However, data provided to me by the National Irrigation Authority (1988) indicate corresponding service areas of 48,355 and 8,386 ha. Once again, the differences are substantial.

These are only two of literally hundreds of examples which could be presented. Contradictions among published data sources in the Philippines are easy to find; one of the primary reasons for choosing 1970 and 1980 as years for analysis is that they correspond to census years.

Since the independent variables to be used in chapter 7 all represent provincial-level data, a major assumption of this study is that interprovincial effects are insignificant. In other words, I assume that forest cover or deforestation is primarily a function of factors originating within a province. While the assumption of zero or minor cross-boundary effects is unrealistic, lack of data on interprovincial interactions makes it necessary. Sungsuwan

(1985) and Panayotou and Sungsuwan (1989) had to make a similar assumption in their study of deforestation in northeast Thailand.

Provincial Data, 1957

The statistical analysis to be conducted on the 1957 data is intended as a preliminary examination. I thought the effort would be worthwhile, since the projected 1957 figure for total forest cover is very close to the 1957 figure presented by the NEC (1959). Thus, I decided to use the 1957 data with the understanding that the data limitations make a direct comparison with 1970 and 1980 impossible.

The NEC (1959) evaluation of forest cover is for midyear 1957 and is based primarily on reports of district foresters. It is not based on remotely sensed data. However, in addition to the fact that their 1957 figure for total forest cover is comparable to the projected figure I derived, another consideration is that since my results were so different from the figures being used by the BF at the time, this may indicate that it was a relatively accurate inventory. At the very least, the analysis of the 1957 data can provide a comparison to the data from later years.

The 1957 data are for fifty-four provinces plus Manila. To account for boundary changes, I combined the provinces of Basilan, Zamboanga del Norte, and Zamboanga del Sur into one province. Manila was dropped, since it had no forest cover. Thus, the total number of provinces is fifty-two.

Forest-cover data are divided into two categories of commercial and noncommercial forest according to the definitions provided in chapter 3. The commercial forest category is further subdivided into accessible and inaccessible forest, with accessible commercial forests defined as "areas which can be profitably exploited with the use of usual logging techniques and facilities" (NEC 1959, 144).

Additional data taken from the NEC (1959) include: area of each province, area of concessions in each province for 1958, and timber production by province for 1957. Although concession area is for 1958, it is assumed that the difference between 1957 and 1958 is not great and is outweighed by having all of the forestry data from one source.

Population data for 1957 do not exist since that year falls between the census years of 1948 and 1960. Hence, population for 1957 was calculated using a straight line projection from 1948 to 1960. The population increase from 1948 to 1957 is the absolute change in population between the 1948 census and the 1957 projected population.

Agricultural data are taken from the Department of Agriculture and Natural Resources (c. 1956, c. 1958, c. 1960). The data were derived from regular surveys conducted by the Bureau of Agricultural Economics. Since there is a great deal of variation from one year to the next with the provincial-level data from these surveys, data on agricultural area are the average of the six years from 1954 to 1959. Some of the shortcomings of these surveys, such as their reliance on the memories of farmers for area sown and the poorer quality of the nonrice data, are discussed by Mangahas (1975) and the Philippine Council for Agriculture and Resources Research (1981). It should be noted that Luning (1981) found the Bureau of Agricultural Economics data to be inadequate and, in some cases, incorrect in his study of the western Visayas (region 6). Unfortunately, these are the only data available.

The distance variable (kilometers) is the straight line distance from the provincial capital to Manila. In a few cases, the largest city rather than the provincial capital was used, since some provincial capitals were quite small. This variable, which will also be used in the 1970 and 1980 data sets, requires an explanation.

The distance variable can capture any of three effects. First, there is a possibility that provinces farther from Manila represent frontier areas where in-migration is occurring and, therefore, deforestation would be expected to be more rapid. Second, distance could be a surrogate for the amount of control (or lack thereof) emanating from Manila and, as a result, deforestation would be expected to be greater, the greater the distance from Manila. Third, if most of the timber harvested had to pass through Manila to be either processed or exported, deforestation would be expected to be greater the closer a province was to Manila, since transportation costs would be less.

This last consideration is unlikely for two reasons. First, although a great deal of Philippine timber has been exported, it is doubtful that many log exports went through Manila, since in most cases they were sent directly from the point of origin to the overseas buyer (Bonita 1977). While there is little evidence on these points, a perusal of the *Philippine Lumberman* for the past forty years indicates that numerous ports, particularly in Mindanao, were used for exporting logs. Second, timber processing mills are located throughout the country (Bonita 1977). In short, if the distance variable captures anything, it is most likely the frontier nature of the more distant provinces and/ or increased ease of avoiding control from Manila.

Road data for 1957 (km) are from the *Yearbook of Philippine Statistics* (Bureau of Census and Statistics 1957). They include national and provincial roads but do not include trails or roads within the boundaries of chartered cities.

Lastly, four variables were constructed from the initial variables: population density (population/province area), agricultural area per person (population/agricultural area), agricultural density (agricultural area/province area), and road density (kilometers of road/province area). The net result is a total of fourteen independent variables that were available for the statistical analysis. The variables for 1957, 1970, and 1980 are listed in table 24.

A drawback of the data to be used for the 1957 analysis is that many of them are not comparable with data which will be used in the cross-sectional examination of forest cover in 1970 and 1980. This is particularly the case with the forestry, agriculture, and population data which, for the latter years, are based on remote sensing and censuses. In addition, there were numerous boundary changes during this period. Therefore a panel analysis between 1957 and 1970–80 will not be attempted.

Provincial Data, 1970 and 1980

Forest Cover

Forest-cover data are of two types: absolute and percentage forest cover for each province. All data are from the P-GFI. Results from the 1980s inventory were compared with the 1969 FRCMs and changes in forest cover were calculated. The only exception to this is the forest-cover data for regions 10 and 11, which were taken from the 1973 *PFS* and represent 1969. One of the advantages of using these data is that all interpretation was done by one group and, therefore, differences in definitions of forest cover and interpretation have been minimized. Forest-cover data represent all forest types.

Another major advantage of the P-GFI is that the forest-cover data for 1969 and the 1980s have been given for present-day provincial boundaries. The study is complete and only the province of Camiguin has been omitted. (Camiguin is 230 km^2 and has been included with Misamis Oriental on Mindanao.) Thus, the data cover seventy-two provinces for the year 1969 and one of the years in the 1980s.

One problem with the data is that forest-cover figures in the 1980s are not all from the same date. Of the seventy-two provinces included, eleven forest inventories were completed in 1980, twenty in 1981, twenty-six in 1984, one in 1985, and fourteen in 1987. Those provinces with inventory end dates in 1981, 1984, 1985, and 1987 were projected back to 1980 using the 1969 to 1981/84/85/87 rate of change calculated on a straight line basis. In addition, all data for 1969 were projected forward to 1970 using the appropriate rate of change for 1969 to 1980/81/84/85/87. All data have been changed to 1970 and 1980 so as to correspond with the population and agricultural census data.

It is important to recognize that the provincial-level data for 1970 and 1980 are for forest cover on public forestlands only. That is, the data do not represent forest cover on A&D lands. This could present three difficulties. First, it means that the forest-cover data understate total forest area. However, this is not considered serious, since forest cover on A&D lands in 1988 was only 1,087 km^2, or 1.7% of all forests (P-GFI 1988). Second, the data understate the rate of deforestation, since deforestation has been more rapid on A&D as opposed to public forestlands (regional reports of the P-GFI 1986 to 1988). Third, it raises the possibility that the change in forest cover between 1970 and

1980 does not accurately reflect deforestation, since public forestlands with forests on them could have been converted to A&D lands. Fortunately, the comparisons made by the P-GFI are between public forested lands in the 1980s and the same land in 1969 on the FRCMs. Regional forest area totals in 1969 and the 1980s for public forestland are almost the same and any changes in total area are slight. In short, even though the data do not capture changes in forests on A&D land and there is a small possibility that some forestland was converted to A&D, these are not considered to be major problems. Lastly, the forest-cover data do not include forest plantations. Since the goal is to understand the process of deforestation, it was not felt appropriate to include them.

Demography

Population. All population data are from the *Philippine Statistical Yearbook* (NEDA 1987). The Philippines conducted censuses in 1948, 1960, 1970, 1975, and 1980. All population censuses were complete enumerations. A census was to have taken place in 1985 but was canceled owing to lack of funds and the social and political unrest which occurred between Senator Aquino's assassination in 1983 and former president Marcos's ouster in 1986. Preliminary results of the 1990 population census are available, but these data are not relevant to the present study.

Urban population. Urban population data for 1970 and 1980 are taken from Special Report no. 4 based on the 1980 census (NCSO 1983). Urban areas for 1970 and 1980 are defined according to one consistent definition and are available for all seventy-three provinces. Urban population data for 1960 are available from Mijares and Nazaret (1974) but have not been incorporated into this study for two reasons: first, the 1960 data is for sixty-six provinces; second, and most important, the 1960 and 1970–80 definitions of urban areas are not the same. The 1960 definition of urban area is based on population density, minimum urban size, and administrative responsibilities, whereas the 1970–80 definition is based on population density combined with urban characteristics. In some cases, the difference between the two is considerable (Mijares and Nazaret 1974).

The rationale for including urban population comes from Luna (1982), who argues that urban centers in the Philippines are strongly correlated with agricultural productivity per worker. Urban centers provide a market for agricultural output, supply needed inputs, and absorb excess agricultural labor.

Upland population. All estimates of upland population come from Cruz et al. (1986). As defined by the Philippine government, upland areas include: (1) lands with slopes of 18% or higher; (2) lands within mountain zones, including tablelands and plateaus at high elevations; (3) lands that are hilly to mountainous in topography (Cruz et al. 1986). According to this definition, approximately 55% of the Philippines is classified as upland.

Using NCSO data and aerial photographs, Cruz et al. (1986) determined that 14.4 million people lived in the uplands in 1980. The potential importance of upland population and its relationship to deforestation flows from the following considerations: (1) technically speaking, almost all lands above 18% slope are reserved as forestlands; (2) since almost all lowland forests are gone, most remaining forests are to be found in the uplands; (3) upland populations have been growing at approximately 2.5% a year; and (4) many observers have commented on the incursion of lowland migrants onto upland forested areas.

Cruz et al. (1986) provide data on upland populations for sixty-seven provinces for the years 1960, 1970, and 1980. They do not present upland population data for Batangas (region 4), Masbate (region 5), Siquijor (region 7), and Basilan and Tawi-Tawi (region 9).

Agriculture. Most agricultural data in the Philippines are collected at the regional or national level; almost all reliable provincial-level data come from the agricultural censuses conducted in conjunction with the population censuses. Even the comprehensive *Data Series on Rice Statistics in the Philippines* (1981) published by the Philippine Council for Agriculture and Resources Research contains primarily regional and national data.

The BAS has done some collecting of provincial statistics since 1960. However, in the process of moving from one office to another during the reorganization brought about by President Aquino's rise to office in 1986, the staff of the research department claim that all provincial statistics from 1960 to 1978 were lost. The BAS claims to have provincial-level data (area and production figures) for rice and some other crops for the period 1979 to present but they are still in the process of "validating" the statistics and, as of August 1988, no provincial-level data on rice or any other crop had been released by the research department of the BAS. The Department of Agriculture in their *Provincial Profiles* (1988) presents provincial-level data for the years 1984–86 on the area of the ten major crops in each province; however, staff of the BAS noted that these data are only provisional, and a comparison with the census data of 1980 revealed large discrepancies between the two data sets. In short, the only reliable provincial-level data on agriculture are from the agricultural censuses of 1960, 1971, and 1980.

The 1971 agricultural census requires an explanation. The actual census was conducted in April 1971 but the data collected concerned the 1970–71 crop year. I have taken the 1971 agricultural census to represent the 1970 crop year.

Data presented in each agricultural census include number and area of farms, area in different crops, and production. The 1971 census was originally for sixty-six provinces but the 1980 agricultural census, which has results for seventy-three provinces, includes the results of the 1971 census for seventy-three provinces also. In other words, no changes in provincial boundaries are necessary for the 1971 agricultural census, since the work has already been done.

While the agricultural censuses are the most complete and, presumably, the most reliable agricultural data in the Philippines, it should be pointed out that Luning (1981) found the 1971 agricultural census to be incorrect in some cases with regard to crop area and farm size in region 6. In addition, Gwyer (1978) notes that it may be the case that the 1971 census did not include approximately 2 million ha of farmland in public forestlands. Unfortunately, this limitation, if true, cannot be overcome.

Infrastructure

Roads. All data on roads come from the research office and library of the Department of Public Highways. They were able to provide me with all of their road data from 1965 to 1987. They have no data before 1965. Since the data leave a lot to be desired, I also made trips to the NEDA (Land Use and Physical Planning Division and Project Monitoring Office), the National Statistics Office, and the Central Bank in an attempt to find other sources of road data. Unfortunately, all data on roads originate from the Department of Public Highways and no agency was able to provide better quality data.

The data for the period 1965–87 include provincial-level data except for the years 1974–81, when they do not exist. No one had a good explanation for this state of affairs but suggestions from government officials included the following: the data were collected but lost; the data were collected but destroyed; or the data were never collected. If the data were never collected, it raises serious questions about the regional figures for 1974 to 1981.

Data are presented for five road systems based on administrative jurisdiction (national, regional, provincial, municipal, barangay) and six types of roads based on construction material (earth, macadam, low-type bituminous, high-type bituminous, concrete, and miscellaneous). According to the NEDA (1981), paved roads account for only 15% of total kilometerage.

Specific problems with the data include: (1) data before 1972 do not include feeder roads but data after 1972 do; (2) it would appear that feeder road and barangay road are interchangeable categories, but this does not hold universally; (3) the categories of barangay, municipal, and city roads are very fluid and it is impossible to tell whether changes in numbers of kilometers from one year to the next are the result of administrative boundary changes, actual changes in roads, or changes in categories; (4) some data are missing; (5) sometimes there are large jumps in the number of kilometers reported from year to year with no explanation given; (6) in some cases, the number of kilometers of roads in a province or region actually declines as one moves forward in time; (7) provincial boundaries are not given, and it is therefore difficult to compare some provinces through time, since their boundaries have changed.

An even greater problem has to do with the actual data. Numerous social scientists in academia and government warned that the data on road networks have to be interpreted carefully. Many researchers doubted that all of the roads on paper actually existed in the field and cautioned that changes in reported kilometers from year to year may represent nothing more than the whims of provincial highway engineers.

Data on roads for 1970 were taken from the information provided by the Department of Public Highways. Since data on roads at the provincial level do not exist for 1979, 1980, and 1981, data were taken from 1982 and projected back to 1980 using the rate of change for 1970 to 1982 calculated on a straight line basis. The data for 1982 are for seventy-three provinces and the data for 1970 are for sixty-seven provinces. I assume that errors in the data are evenly distributed across all provinces and, therefore, accurately represent the relative extent of road networks in each province in 1970 and 1980 and that changes in road data from 1970 to 1980 accurately reflect actual changes in road networks. Even if the absolute levels of road networks are not accurate, I assume that relative levels and changes in those levels are.

An important caveat is in order: data on logging roads do not exist. While some logging roads eventually become barangay or municipal roads, there are no data regarding extent and location of logging roads. Some of the larger concessions have accurate maps of their roads but most logging roads (legal and illegal) are not recorded. In general, the category of logging roads is fluid, as new ones are built, old ones are abandoned, and others are taken over by local governments. In addition, logging roads can vary from little more than dirt tracks (or stream beds, in the case of some concessionaires) to improved, all-weather roads.

Since data on this important category of roads are not available, I assume that the total of national, provincial, municipal, city, and barangay roads in a province is an accurate indication of the relative extent of all road networks across all provinces.

Irrigation. All irrigation data come from the corporate planning office of the National Irrigation Authority (NIA). Their data are for three types of irrigation systems: national systems, which are built and run by the NIA; communal systems, which are built by the NIA but operated by local irrigation groups; and pumps and private systems, which are tubewell irrigation systems built and run by private individuals.

Data for all three systems are in hectares and represent the service area covered by each system. Data cover the years 1965–87 and are presented for all seventy-three provinces so that adjustments in provincial boundaries were not necessary. Data from all three systems were summed so that irrigated area represents the service area of all three systems.

Physical Variables

The physical variables are province area and arable land per province. Provincial area comes from Special Report no. 3 of the *1980 Census of Agriculture* (NCSO, n.d.) and provides data for all seventy-three provinces; thus, boundary adjustments were not necessary.

Land capability classes are taken from the Bureau of Soils (1977). Arable land is considered to be land suitable for annual crops. It is not a measure of actual land use but of land suitability. It is to be expected that deforestation would be more rapid in provinces with larger areas of arable land. Because data were available for only forty-nine provinces, numerous adjustments had to be made in order to render them compatible with the present seventy-three provinces.

Forestry sector

Logging data consist of statistics on the number, size (in hectares), and ACC (in cubic meters) of concessions and actual production (in cubic meters) at the provincial level. All data are taken from the *PFS*. A major problem with the *PFS* data is that there are large gaps in the figures presented from year to year. For instance, provincial-level data on logging production do not exist for the years 1975–82, and provincial-level data regarding concessions do not exist before 1973. In addition, the FMB does not have the necessary background data to reconstruct logging and concession data for these years. A conversation with a government official in 1988 revealed that when they moved offices in 1987 at least 10% of all their files were "lost." The lost files included the data that are of most interest to this study.

Concessions: Number, area, and ACC. Since statistics on the number, size, and AAC of concessions are available for the years 1972 and 1980 on a provincial basis, I used the 1972 data in place of any from 1970. Adjustments in the data for 1970 were necessary because of changes in provincial boundaries.

Logging production. Log-production figures for fiscal years 1969–70 and 1970–71 are presented in the *PFS* for 1972. The two fiscal years were averaged to obtain a figure for 1970. Since only regional production figures are available for the years from 1975 to 1982, the 1980 production data represent 1983 provincial figures adjusted to the 1980 regional totals. Adjustments in the 1970 data were necessary because of changes in provincial boundaries.

Variables not included

Climate

The nature of monsoonal (seasonal) forests has been suggested as a factor influencing the rate of forest depletion (Whitmore 1984). Since monsoonal forests in the Philippines experience a period of little rain for approximately four to six months a year, it might be expected that they would have higher rates of deforestation, other factors remaining constant, due to their increased accessibility and greater susceptibility to burning (Olofson 1981). In a similar vein, Lugo et al. (1981) in their study on deforestation in the Greater Caribbean, suggest that wet life zones are the last to lose their forest cover because their climates are "too wet" for agriculture. Unfortunately, I was not able to test this proposition for several reasons.

First, several climate classification schemes based on rainfall exist for the Philippines and the maps based on these criteria often disagree (see Huke 1963, 44–51; Huke 1982; Kintanar 1984, 29–37; National Water Resources Council 1977, 4 and 7; Wernstedt and Simkins 1967, 58–62). Second, regardless of the climate classification scheme used, quite a few provinces partake of more than one climate category (Department of Agriculture 1988) and it is impossible to assign each province to a unique classification. Lastly, there is no historical or contemporary provincial-level data mapping the relationship between climate and forest cover.

Government Resettlement Projects

Government-initiated land settlement projects in the Philippines can be traced to the early twentieth century (Paderanga 1986). According to data from the Department of Agrarian Reform (1987), the agency currently responsible for settlement projects, there were forty-six resettlement sites in 1987 covering approximately 600,000 ha and including 64,000 families. Twelve sites are on Luzon, one on Palawan, nine in the Visayas, and twenty-four on Mindanao.

According to the Institute of Population Studies (1981), settlers are classified into one of four categories: pioneer settlers are those who moved to the site before the project began; moved-in settlers were actually assisted by the government; self-propelled are spontaneous settlers who moved in without government assistance; and local settlers lived in the project site before it began and changed to settler status at a later date. In addition, a new category of rebel returnees (Communist and Muslim) has been created (Department of Agrarian Reform 1988).

Government resettlement projects as an explanatory variable will not be considered for three reasons. First, the numbers involved are small in relation to internal population movements. Second, data on the arrival dates of settlers are not available. Third, given the multiplicity of settler categories, it is very difficult to determine how many settlers were moved in by the government. Bahrein (1988) argues that only 28% of all settlers are actually assisted, and Wernstedt and Spencer (1965) feel that for Mindanao 10% would be a generous estimate. Regardless of the exact number, it would appear that the number of spontaneous settlers is much larger than that of moved-in settlers (Abad 1981).

Fuelwood

According to a 1970 government survey, 79% of all Philippine households used wood as a source of fuel (Hyman 1983; Tesoro 1989). In addition, some cottage industries such as tobacco curing are dependent on fuelwood (Hyman 1983, 1984). Overall, the quantity of wood used as fuel is significantly greater than legal log production (Hyman 1983; Tesoro 1989). However, a variable for fuelwood use has not been included for two reasons.

First, provincial-level data on fuelwood use do not exist except for a few case studies, and fuelwood statistics in general are of doubtful accuracy (World Bank, 1989a) (see Metz 1989 and Ives and Pitt 1988 for a discussion of the problems of fuelwood statistics in Nepal). Second, and most important, there is little evidence that deforestation in the Philippines, or any other Southeast Asian country, results from the demand for fuelwood (Gillis 1988). Fuelwood is obtained by lopping off branches of large trees, cutting small trees in brushland (not forestland), cutting trees which have been grown for the specific purpose of providing fuelwood, or gathering after forests have been cleared for agriculture (Leach and Mearns 1988; Marx 1985; Tesoro 1989; personal observation of author). While charcoal making can directly cause deforestation (Huke 1986; personal observation of author), fewer than 0.5% of all Philippine households in 1970 used charcoal for cooking (NEDA 1980). In addition, no statistics exist regarding charcoal use at the provincial level.

Migration

The National Statistics Office (formerly the NCSO) has migration data at the provincial level for 1960–70, 1970–75, and 1975–80. However, these data will not be used. The 1970–75 data seriously understate the extent of migration (possibly by as much as 30%) and the 1975–80 data understate migration by 5–10% (Zosa-Feranil 1988; Nguiagain 1985). In lieu of migration data, changes in total provincial population as indicated by the postwar censuses (1948, 1960, 1970, 1980) will be used. Thus, population change from one census year to the next will represent the sum total of the natural rate of increase (crude birthrate minus crude deathrate) and net migration.

Poverty

Data on poverty, per capita income, distribution of income or assets, landlessness, and unemployment at the provincial level are almost completely lacking. In fact, several studies have raised serious questions about the reliability of data on these variables at the national level (APST 1986; DeDios 1984; Dowling and Soo 1983; Feder 1983). Under these circumstances, it was not possible to include any of these socioeconomic variables.

Commercial Agriculture

It was previously noted that commercial or export agriculture could result in deforestation in either of two ways: directly, through expansion of land under crops, or indirectly, by pushing poor people off nonforestland who then must practice subsistence agriculture elsewhere. In the Philippines, these arguments have been made by Feder (1983), Ofreno (1980), Tiglao (1981), and Tiongzon et al. (1986). However, it was not possible to include such a variable for two reasons.

First, the terms used in discussions of this topic are vague. If "commercial agriculture" means producing for the market, then most rice production would have to be included; at the same time, rice production cannot be considered production for export, since very little, if any, rice is exported from the Philippines. Additional terms found in the literature include "agri-business," "corporate farming," and "plantation agriculture." Unfortunately, these terms have little operational meaning and are rarely defined in a manner which would facilitate quantitative analysis.

Second, coconuts, which comprise about 75% of all "commercial" cropland (Tiglao 1981), are actually a smallholder crop with an average farm size of 3.4 ha (Ofreno 1980). Approximately 75% of all coconut farms are less than 5 ha (Sacerdoti and Galang 1985). Sugar, the next largest commercial crop in area, is exported, but about 50% of all production is consumed in the Philippines (David and Barker 1979; Sacerdoti and Galang 1985).

Given the above considerations, I chose two variables for agricultural area: total agricultural area and the combined area of rice and corn farms. Rice and corn account for approximately 80% of total staples consumed (Bennagen 1982) and will be used as a measure of food availability. The available data do not permit an examination of the relationship, if any, between commercial/export agriculture and deforestation.

Other Variables

This study, unlike the work of Sungsuwan (1985) and Panayotou and Sungsuwan (1989) for northeast Thailand, does not include any variables for the prices of crops, fuelwood, charcoal, or nonwood fuels. The reason is that provincial-level data on these variables do not exist.

Cattle ranching has not been included because it is not important in the Philippines or in Southeast Asia (Gillis 1988). In addition, even though grasslands exist throughout the Philippines, no one has ever drawn a correlation between deforestation and cattle ranching (in contrast to the Brazilian Amazon, for example). What cattle ranching does exist occurs in grasslands that followed shifting cultivation that followed logging (Segura-de los Angeles 1985).

Lastly, peace and order considerations may be important in some areas; unfortunately, there are no accurate data on this topic. There have been two major armed struggles in the postwar period: the Muslim separatist movement and the Communist rebellion led by the New People's Army (Costello

1984; Gowing 1980). The fighting in Mindanao between government and Muslim troops during the mid-1970s was intense (Molloy 1983) and may account for the substantial decreases in population in Lanao del Sur between 1970 and 1980, and in Sulu between 1970 and 1975, and the almost negligible population increase in Maguindanao and north Cotobato between 1970 and 1975. In addition, Nguiagain (1985) reports that out-migration from Luzon after 1970 shifted from Mindanao to the Visayas and claims that some of this was the result of the unrest in Mindanao. The ultimate effect of these population movements and the fighting on logging and deforestation is unknown.

Summary

Most of the data presented in this study were gathered during a ten-month stay in the Philippines from 1987 to 1988. I collected almost all data from national government agencies in Manila or the various forestry organizations in Los Baños. I made no attempt to gather data directly from provincial governments, because visits to all seventy-three provinces would have been impractical; and on the basis of personal experience and the opinions of numerous social scientists (Filipino and foreign), I determined that it would not be worth the effort.

The cross-sectional dates of 1957, 1970, and 1980 have been chosen to correspond as closely as possible to the available forestry and census data at the provincial level. Thus, to a large extent, the availability of data has determined the structure of this study.

Because of data limitations, many of the factors presented in the flow diagram (fig. 2) cannot be included in the statistical analysis. These would require direct measures of elite control of government and corruption, of ownership of concessions, and of poverty and landlessness. In addition, given the widespread existence of illegal logging and corruption in the forestry sector, there are reasons for doubting that the logging data as reported by the BFD are reliable. As Hicks and McNicoll (1971, 212) note: "Statistics based on legal logging are . . . quite inadequate as a measure of the rate of decline of forest area."

Unfortunately, it is not possible to overcome the limitations of the logging data. Even if it was known with certainty, for example, that logging was twice as great at the national level as actually reported, this knowledge would not be of help. Without detailed information as to where the logging occurred, multiplying all logging figures by 2 does not provide any additional information, since it is simply a linear transformation of the provincial data. It is most likely that accurate logging figures at the provincial level for the postwar Philippines will never be available.

7

Statistical Analysis of Deforestation

Introduction

The main interest of this study is deforestation; however the cross-sectional analyses of 1957, 1970, and 1980 are not concerned with deforestation per se; rather they deal with the relationship of absolute forest cover across provinces at one point in time with the hypothesized independent variables. In contrast, the panel analysis deals with the absolute change in the independent variables from 1970 to 1980 and, as such, is directly concerned with deforestation, since the dependent variable is the absolute change in forest cover. The major difference between time series analysis and panel analysis is that in panel analysis the observations are taken at relatively few points in time (in the present case, only twice) (Markus 1979). The discussion in chapter 5 and the flow diagram depicted in figure 2 provide the background to this statistical analysis. The variables for which data are available that would be expected to be significantly related to forest cover are agricultural area, roads, and population. In the cross-sectional analysis, logging activity, concession area, and ACC are not considered to be explanatory variables of forest cover; rather, the relationship is assumed to be just the opposite: that is, forest cover determines the amount of logging, size and number of concessions, and AAC.

The consideration of two dependent variables for the cross-sectional analysis (absolute and percentage forest cover) requires further explanation. The reason for including absolute forest cover is premised on the notion that commercial loggers are primarily motivated by the size and nature of the available forest; this would be the case particularly for those concessionaires who made long-term investments in plywood and veneer plants. Loggers are concerned with cubic meters of commercial timber per hectare and the total number of hectares of forest cover. At the same time, potential migrants are

Table 23.
Level of explanation (r^2) from regressing timber production and concession size on absolute and percentage forest area

Dependent variables		Independent variables FA	%FA
Logging	1957	0.29	0.21
	1970	0.28	0.23
	1980	0.34	0.28
Concession size	1957	0.73	0.55
	1970	0.71	0.41
	1980	0.53	0.25

Note: The two dependent variables are annual production of timber in cubic meters and concession size in hectares, and the two independent variables are absolute forest area (FA) in hectares and percentage forest area (%FA).

more likely to be aware of recently logged areas rather than the percentage forest cover of a province, since logged forests or open land represent subsistence opportunities for poor people.

Table 23 indicates that absolute rather than percentage forest area is more important in explaining the distribution of legal logging and concession size. In every case the F and t-statistics are greater for forest area. The lower r^2 for logging as compared to concession size most likely indicates the widespread nature of illegal logging and overcutting by legal concessionaires, neither of which is captured in the official logging data. The declining r^2 for concession size from 1957 to 1980 most likely represents the fact that as forest area decreased, more and more concessions came to include larger and larger nonforest areas within their boundaries.

One of the reason for using percentage forest cover as an independent variable is that absolute forest cover and provincial area are highly correlated. The Pearson correlation coefficients for provincial area and absolute forest cover in 1957, 1970, and 1980 are .87, .84, and .80 respectively. Using percentage forest cover is an attempt to control this strong correlation. The Pearson correlation coefficients for provincial area and percentage forest cover in 1957, 1970, and 1980 are .49, .41, and .37 respectively.

Regression analysis that used percentage forest cover as the dependent variable did not yield satisfactory results as compared with using absolute forest cover. In general, r^2s were lower and heteroscedasticity and/or serial correlation was a problem. Since the major factor in the decline of primary forest in the postwar Philippines has been logging and the expansion of agriculture in degraded forests, I conclude that absolute as opposed to percentage forest cover better fits the model of deforestation presented in chapter 5. An additional consideration is that logging concessions are granted on the basis of forest area regardless of provincial boundaries. As an example, the concession of Paper Industries Corporation of the Philippines (PICOP) in eastern Mindanao covers approximately 170,000 ha and parts of four provinces (PICOP 1988).

In short, this study, in contrast to most of the quantitative work described in chapter 2, will use absolute rather than percentage forest cover as the dependent variable in the cross-sectional analyses. The motivation for loggers is absolute forest area, and for new migrants it is the amount of degraded forest. It is not at all clear that percentage forest cover is a motivating factor for the agents of deforestation.

Province area as an independent variable is not included, for several reasons. First, as the high Pearson correlation coefficients between province area and forest cover indicate, forest cover is strongly related to province size. The addition of province area does not further our understanding of the variation in forest cover as the result of socioeconomic processes. The contribution of province area is obvious. Second, the inclusion of province area introduced a great deal of heteroscedasticity to the estimated equations. This flows from the large range of province areas. In 1957, the range was from 198 to 22,968 km^2 and in 1970–80 it was from 209 to 14,896 km^2. Third, because the number of provinces in 1957 was fifty-four (reduced to fifty-two) and in 1970 and 1980 was seventy-three (reduced to sixty-eight and seventy-two respectively), the variable is not consistent over the different cross-sections.

Lastly, given that my subject is deforestation, an explanation is in order regarding why cross-sectional analyses were conducted. First, all previous quantitative work on tropical deforestation has been cross-sectional in nature, and such an analysis is useful for purposes of comparison. Second, since data are available for three years (1957, 1970, and 1980), I decided that cross-sectional analysis from three widely separated dates might provide information on the changing nature of the deforestation process over time. However, I have serious doubts as to the ultimate usefulness of cross-sectional analysis in articulating the dynamic nature of the deforestation process and, additionally, feel that the real test of the deforestation model presented in chapter 5 is the panel analysis which follows the cross-sectional analysis.

Table 24 lists the variables which were available for each cross-sectional analysis. The 1970–80 panel analysis included the absolute changes of these same variables. The sample size is fifty-two for 1957, sixty-eight for 1970, seventy-two for 1980, and sixty-seven for the panel analysis. Four provinces could not be included in the 1970 cross-sectional analysis because they had no forest cover and could not be logarithmically transformed. As mentioned previously, owing to the incompatibility of some of the data between 1957 and the later years, a panel analysis was not conducted using that year.

In general, the statistical analysis was carried out using the following steps: (1) variables were plotted to see if any patterns or clustering were apparent; (2) Pearson correlation coefficients among all variables listed in table 24 were determined; (3) multiple regressions of the independent variables de-

Table 24.
Variables available for the statistical analysis at the provincial level

	1957	1970 and 1980
Area (km^2)	X	X
Arable land (km^2)		X
Total population (1,000s)	X	X
Urban population (1,000s)		X
Upland population (1,000s)		X
Increase in total population (1,000s)	X	X
Increase in upland population (1,000s)		X
Population density (per km^2)	X	X
Upland population density (per km^2)		X
Total agricultural area (km^2)	X	X
Rice and corn area (km^2)	X	X
Rice and corn production (tons)		X
Agricultural area per capita (ha)	X	X
Rice and corn productivity (tons/ha)		X
Percentage agricultural area	X	X
Rice and corn production per capita		X
Number of farms (1,000s)		X
Irrigated area (km^2)		X
Irrigation density (per km^2)		X
Roads (km^2)	X	X
Road density (per km^2)	X	X
Logging (m^3)	X	X
Number of concessions	X	X
Area of concessions (ha)	X	X
AAC (m^3)		X
Distance from Manila (km)	X	X
Total number of variables	14	26

termined to be important on an a priori basis were run against absolute forest cover; (4) stepwise regressions were run to see if any of the other variables were important; (5) the resulting equations were standardized; (6) the residuals were plotted and tests for heteroscedasticity and serial correlation were performed; (7) if the resulting equations had problems with heteroscedasticity and/or serial correlation, the variables were transformed logarithmically and reestimated.

The level of significance has been set at 5% and all variables included in the equations meet this standard. In the tables which follow, all coefficients have been standardized and t-statistics are presented in parentheses below the corresponding variables. Adjusted r^2, F, chi-square (test for heteroscedasticity, where p greater than .05 indicates that the results are significant at the .05 level), and Durbin-Watson (test for serial correlation) statistics are to the right of each equation.

Statistical Results for 1957, 1970, and 1980

Table 25 presents the multiple regression results for 1957, 1970, and 1980. Both dependent and independent variables are in natural logarithms for the years 1957 and 1970. Since both equations are standardized and in double log form (the L in front of the variable name indicates that it is a logged variable), the beta coefficients represent the responsiveness of the dependent variable with respect to each of the separate independent variables. The equation for 1980 is not in log form. T-statistics are in parentheses below the appropriate variable.

According to table 25, forest area in 1957 is negatively related to road density (RdK). The sign of the variable is as expected and road density accounts for approximately 75% of the variation in provincial forest cover. The interpretation of the beta coefficients is straightforward; for example, equation 1 indicates that a 10% increase in road density leads to an 8.7% decrease in forest cover. In short, forest cover was most extensive where road networks were least dense.

A potential problem with the data has to do with correlations among forest area (FA), population density (PopK), road density (RdK), and the percentage of provincial area under agriculture (%AgA). As an example, the correlations among these variables for 1957 are presented in table 26.

While it is obvious that the correlation between forest area (FA) and road density (RdK) is stronger than with the other two variables, strong correlations still exist among the three independent variables. This indicates that forest cover is negatively related to road, population and agricultural density. Unfortunately, it is conceptually difficult to delineate clearly the relationships among these three independent variables. While roads open up forest areas

Table 25.
Multiple regression results for 1957, 1970, and 1980 showing the importance of road density (RdK), population density (PopK), and the change in roads from 1970 to 1980 (Rd7d8)

1957

(1) $L(FA57) = -0.87\ L(RdK)$ $r^2 = 0.76$, F = 159
    N = 52 (-12.6) Chi-square = 4.69
                                                            DW = 1.70

1970

(2) $L(FA70) = -0.54\ L(PopK) - 0.28\ L(RdK)$ $r^2 = 0.58$, F = 48
    N = 68 (-4.7) (-2.4) Chi-square = 11.0
                                                                          DW = 1.66

1980

(3) $FA80 = 0.32\ Rd7d8 - 0.38\ RdK - 0.29\ PopK$ $r^2 = 0.41$, F = 17
    N = 72 (3.5) (-3.5) (-2.6) Chi-square = 10.9
                                                                                  DW = 1.92

Table 26.
Correlations among forest area (FA), population density (PopK), road density (RdK), and percentage agricultural area (%AgA), 1957

	PopK	RdK	%AgA
FA	− 0.68	− 0.87	− 0.68
PopK	—	0.82	0.88
RdK	—	—	0.78

to loggers and settlers follow the loggers, this general statement needs qualification.

First, we have no way of knowing whether, in fact, the road data include logging roads. In other words, it is not clear that the road data include those roads most likely to lead to deforestation. Second, it may be the case that roads are not necessary for the establishment of agriculture or human settlements in some upland/forest areas. Spencer (1966) reports that in Mindanao in the 1950s, settlers followed loggers and, at times, preceded them, and Lopez (1989) has observed this same phenomenon in the frontier areas of Palawan. In addition, for those ethnic groups who have traditionally lived in the uplands, human settlements and the resulting deforestation have existed for decades without the construction of road networks. These ethnic groups (excluding Muslims) numbered about 3.5 million in 1980 and represented approximately 7% of the Philippine population (Republic of the Philippines/ UNICEF 1987). Third, it may be that while large-scale roads are necessary for the creation of new human settlements (see Vandermeer and Agaloos 1962 and Simkins and Wernstedt 1971 regarding Mindanao), smaller roads are not as important for the deforestation process.

In short, the data problems discussed above and the almost complete lack of detailed case studies regarding changing land use in the Philippines restrict, to a certain extent, our ability to interpret the statistical results. Under these circumstances, it may be more appropriate to view road density as a proxy for the entire process of human settlement, at least for 1957. To test this hypothesis, a principal components analysis (PCA) was done using eleven of the independent variables; namely, population (Pop), population density (PopK), roads (Rd), road density (RdK), distance from Manila (Dist), agricultural area (AgA), percentage agricultural land (%AgA), area of rice and corn (RCAr), agricultural area per person (AgAP), increase in population (PopIn), and provincial area (Area).

Since Batanes province had out-migration, it was dropped, because the log transformation cannot handle negative values. (Batanes is 198 km^2 and represents only 0.0007% of the total land area of the Philippines.) Thus, the sample size for the PCA is fifty-one. In addition, since there are only eleven independent variables, the number of factors was arbitrarily set at two. The factor loadings and the corresponding variables are in table 27.

Table 27.
Factor loadings from the principal components analysis on the 1957 data set (in natural logarithms)

	Factor 1	Factor 2
LPop	0.93	0.15
LRd	0.90	0.01
LDist	0.21	-0.60
L%AgA	0.45	0.78
LPopK	0.35	0.92
LRdK	0.09	0.95
LAgA	0.96	-0.10
LRCAr	0.94	-0.13
LAgAP	-0.12	0.45
LArea	0.46	-0.85
LPopIn	0.66	-0.17

Factor 1 represents the sum total of activity associated with human settlements (roads, population, and agricultural area) and will be called the settlement factor. Factor 2, on the other hand, represents a density and size variable. If these factors are considered as independent variables they can be regressed against the logarithm of forest area. The resulting equation yields an r^2 of 0.71 with t-statistics for factors 1 and 2 of 1.1 and –11.1 respectively. In other words, factor 1 is not a significant explanatory variable with regard to the variation in provincial forest cover.

The high factor loadings of road, population, and agricultural density with factor 2 support the previous discussion regarding the results of the multiple regression. The use of the road density variable in the multiple regression appears to be capturing the combination of factors represented by human settlements. A tentative conclusion is that the 1957 data represent an early stage of pioneer settlement, with forest cover negatively related to human settlements.

The cross-sectional analyses for 1970 and 1980 were conducted in a manner similar to that for 1957. The number of variables available for the analysis was much greater (n = 26) but this did not have any appreciable effect on the resulting equations: road density (RdK), population density (PopK), and changes in roads from 1970 to 1980 (Rd7d8) were the most important variables. Similar difficulties were encountered with high correlations among forest area, agricultural area, and road/population densities; to compensate, a PCA was also conducted for each of these years.

Table 25 indicates that forest cover in 1970 is negatively related to road and population density. The results for 1980 are very similar to those for 1970. The major difference is that the results for 1980 indicate that the increase in roads from 1970 to 1980 is important and may indicate continued movement to frontier/upland areas. The signs of the density variables are as expected.

In table 25, road density appears in all three equations, population density in two, and the increase in roads in one. Agricultural density does not appear in any of the equations. In addition to the fact that the variables are not the same for all three equations, it is of interest to note that the r^2 declined from 0.76 in 1957 to 0.58 in 1970 and 0.41 in 1980. These results may indicate the changing nature of the forces correlated with forest cover. Seeing that the correlations among population, road, and agricultural densities and forest cover for both 1970 and 1980 were high, I conducted a PCA and regressed the resulting factors against forest area. The number of factors was arbitrarily set at five.

For 1970, the only factor that was significant was factor 2 and it represents a density factor (high factor loadings with agricultural, road, and population density). When regressed against forest area it yielded an r^2 of 0.68. For 1980, three factors were statistically significant: factor 1 (the settlement factor), factor 2 (density), and factor 3 (distance from Manila and agricultural density). These three factors yielded an r^2 of 0.60.

The PCA of 1970 is almost exactly the same as the PCA of 1957; in both cases road, population, and agricultural densities (factor 2) are the most important. On the other hand, the PCA for 1980 is different from the other two because, in addition to factor 2, factors 1 and 3 were significant. Given that the deforestation process was well advanced by 1980, this finding is not surprising and will be discussed in conjunction with the panel analysis below. At this point, we can simply note that the differences among the three cross-sectional analyses appear to be capturing different phases of the history of the forest cover in the postwar Philippines.

The importance of population density in 1970 and 1980 and the relative decline of the importance of road density after 1957 may indicate that a fundamental change occurred between 1957 and 1980. As mentioned above, the 1957 data appear to be capturing the initial stage of frontier settlement. A plausible interpretation is that large-scale road projects were opening up previously inaccessible areas of the country, particularly Mindanao (Vandermeer and Agaloos 1962; Wernstedt and Simkins 1965).

On the other hand, the increasing relative importance of the population density variable as indicated by its greater correlation with forest area, its appearance in the 1970 and 1980 equations, and its greater beta coefficient in 1970 demonstrates that road density was no longer playing as pivotal a role as before. Once again, this demonstrates the role that roads play in opening up an area.

It must also be kept in mind that the negative relationship with population density does not mean that no population growth was occurring in those provinces with larger forest areas. It means, instead, that population growth was greater in those provinces with less forest area. In short, it supports the view that some time during the 1970s there was a shift in internal migration away from the frontier areas to the more urbanized provinces (especially the Metro Manila Area).

The cross-sectional analysis confirms the analysis presented in chapter 5 with one exception: agricultural area does not appear to be significant. A possible explanation for this has been suggested earlier: the questionable nature of some of the agricultural data. This would be particularly the case with the 1957 data, which are not based on an agricultural census. At the same time, Luning's experience with the 1971 agricultural census (1981), indicates that even census data may be weak at enumerating agricultural expansion in frontier/upland areas.

Panel Analysis, 1970–80

In the panel analysis, the dependent variable is the absolute change in forest cover between 1970 and 1980. This variable is a direct measure of deforestation. The independent variables primarily represent changes in the variables used in the 1970 and 1980 cross-sectional analyses. The sample size is sixty-seven. Four provinces were not included because they had no forest cover in 1970. An additional province was not included because it had a slight increase in forest cover and it is doubtful that this is an accurate reflection of the actual change in forest cover over this period. These five excluded provinces total 6,499 km^2, or 2.17% of the national land area.

In the panel analysis, forest area in 1970 is not included as an independent variable. The Pearson correlation coefficient between deforestation from 1970 to 1980 and forest cover in 1970 is 0.73. In addition, a regression of deforestation as a function of forest area in 1970 yields an r^2 of 0.53 and the result is statistically significant at the 0.0001 level. In short, deforestation occurs primarily where forest cover is the greatest. This result is obvious and does not explain the socioeconomic forces causing deforestation. As such, forest area in 1970 will not be considered as an independent variable for further analysis.

Based on the flow diagram in chapter 5, the variables for which data exist that should be important are the change in agricultural area, logging in 1970, change in population, and distance from Manila (a measure of corruption or lack of control of the forest removal process). Since logging in the postwar period is considered to be strongly underreported, I decided to use data on the AAC instead. AAC data (in cubic meters) is available on a provincial basis for 1972 and represents what the BF determined was an appropriate yield for the concessions in a particular province. I assume that the difference between 1972 and 1970 is not significant. In the postwar period, AAC has been greater than legally reported logging for almost every year. This is a reflection of the underreporting of timber extraction by logging firms. Since the AAC figures are greater than the logging figures, they should more accurately reflect actual timber harvesting. Table 28 presents the correlations among those variables determined to be important on an a priori basis.

Table 28.
Correlations among deforestation (Def), forest area in 1970 (FA70), change in agricultural area
(DAgAr), change in population (DPop), distance from Manila (Dist), and AAC in 1970 (AAC70)

	Def	FA70	DAgAr	DPop	Dist	AAC70
Def	1.00					
FA70	0.73	1.00				
DAgAr	0.59	0.41	1.00			
DPop	0.26	0.05	0.20	1.00		
Dist	0.44	0.26	0.42	0.03	1.00	
AAC70	0.61	0.69	0.39	0.18	0.47	1.00

The initial equation estimated was deforestation in the 1970 to 1980 period (Def) as a function of distance from Manila, change in agricultural land,
change in population, and ACC in 1970. In the initial equation, neither the
change in population nor distance from Manila was significant; thus, distance
was dropped as an independent variable and the equation reestimated. However, the change in population was not significant in the reestimated equation, so that the final equation included only the change in agricultural land
and the ACC in 1970 as independent variables. Even though the sign for the
distance variable was positive as expected, it was not statistically significant.
This is a reflection of the high multicollinearity between distance and change
in agricultural area ($r = 0.42$) and ACC ($r = 0.47$). Deforestation as a function of
distance alone yields an r^2 of 0.18 and the result is statistically significant. The
results of the panel analysis are presented in standardized form in table 29. T-
statistics are in parentheses below the corresponding beta coefficients.

Table 29.
Results of the panel analysis (1970–80) showing the importance of the change in agricultural
area (DAgAr) and AAC in 1970 (AAC70)

Independent variables	(1)	(2)	(3)
Dist	0.10	—	—
	(0.94)	—	—
DPop	0.11	0.09	—
	(1.18)	(1.08)	—
DAgAr	0.37	0.40	0.41
	(3.69)	(4.20)	(4.40)
AAC70	0.40	0.44	0.41
	(3.90)	(4.60)	(4.79)
F statistic	18.00	23.00	34.00
DW statistic	1.93	1.98	1.99
Alpha*	0.08	0.17	0.06
Adjusted r^2	0.50	0.50	0.50

*Alpha is the level of significance for the chi-square statistic which measures heteroscedastic-
ity. An alpha greater than 0.05 indicates that heteroscedasticity is not a problem at the 0.05
level of significance.

Following Kennedy (1979), a test for multicollinearity was conducted using condition indices. The highest condition index of 6.3 is below the level of 30 (Kennedy 1979), which indicates that collinearity is not a problem. In addition, when the independent variables in equation 3 are regressed against each other, the r^2 obtained is only 0.14, which is considerably less than the 0.50 obtained in the equation for deforestation, and this also indicates that multicollinearity is not a serious problem (Kennedy 1979). In short, there would appear to be no major statistical problems with equation 3.

Table 29 indicates that deforestation in the 1970–80 period is positively related to ACC in 1970 and changes in the area under agriculture. It is important to remember that ACC is being used as a proxy for logging, and that both logging and ACC, to a certain extent, represent the corruption factor.

In chapter 6, it was suggested that the distance variable represented the degree of control or, more properly, the lack of control that government officials in Manila were able to exercise over use of the forests, particularly logging and migration by poor farmers to previously forested areas. The importance of the distance variable was indicated by the regression analysis between the distance variable and deforestation. In addition, it was further confirmed by the PCA for 1980. In the panel analysis (1970–80), deforestation was found to be positively correlated with distance from Manila and the ACC in 1970. It seems that the model presented in chapter 5 has been confirmed. In addition, given the inaccuracy of the logging data, the relationship between logging and deforestation would be even stronger if accurate data were available.

Although the details of who received logging concessions, the extent of illegal logging, and the role of the military in protecting these activities will never be known with certainty, it is the opinion of most observers that under President Marcos awards of logging concessions were used to retain the loyalty of friends and the military. I do not consider it a coincidence that the highest rates of deforestation occurred during the 1970s when the Philippines was under martial law (see table 9).

The above discussion leads to an important qualification of the interpretation of the distance variable. While it has been suggested that it represents lack of control from Manila, it is most likely more appropriate to say that it represents lack of control by the agency ostensibly in charge of forests, the BF/BFD. In actuality it meant that control of the forests was placed in the hands of private interests. In other words, the process of awarding concessions and protecting those engaged in illegal logging was actually an orderly process with the benefits flowing to a relatively small group of people. It is interesting that in those areas of the Philippines where logging was banned during the Marcos era, rates of deforestation were actually higher than where logging was allowed (Schade 1988). While the process of deforestation may appear to be chaotic, enough order has existed to ensure that the benefits of the process have flowed to those in positions of power.

The positive relationship between the spread of agriculture and defor-estation in the panel analysis confirms the model of deforestation presented in chapter 6. In addition, the importance of the spread of agriculture is sup-ported by the 1980 PCA. Of interest is the fact that when the change in popula-tion from 1970 to 1980 is regressed against deforestation, the r^2 is only .05. When the change in population density is regressed against deforestation, the r^2 is only .02 and the result is not statistically significant. In short, the results of this study cannot support the contention of most scholars of tropical defor-estation that increasing population is the leading cause of deforestation.

A caveat is in order regarding the significance of the change in agricul-tural area in the panel analysis. The discussion in chapters 4 and 6 revealed that the 1971 *Census of Agriculture* may have undercounted the extent of farmland, particularly in upland areas. If this is so and if the 1980 *Census of Agriculture* was accurate, then the change in agricultural area from 1970 to 1980 would have been exaggerated and this may account for its appearance in the panel analysis. Since the accuracy of the 1971 *Census of Agriculture* is still an open question (Gwyer 1978), this issue cannot be answered with certainty.

The r^2 of the panel analysis is 0.50 and the r^2 of the cross-sectional anal-ysis ranges from 0.41 to 0.76. While these results are encouraging, it is appro-priate to ask what factors may prevent these r^2s from being even higher. Sev-eral explanations suggest themselves. First, it may simply be the case that the data, or, at least some of them, are inadequate for the task at hand. The lack of accurate information on logging is especially important in this regard. Sec-ond, certain variables were not included because data simply do not exist, al-though they were determined on an a priori basis to be relevant, for example,

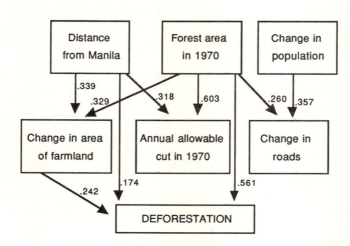

Fig. 3. Path analysis (with forest cover)

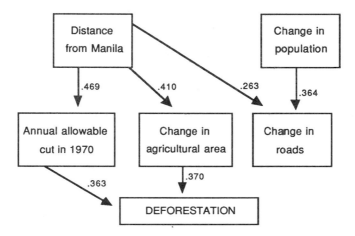

Fig. 4. Path analysis (without forest cover)

poverty and landlessness. Third, the deforestation process may be much morecomplex than hypothesized and caused by factors not included—for example, an export ban on wood products by another tropical country. Given the lack of detailed case studies, model uncertainty must be considered to be a possibility. Lastly, it could be a combination of all three of the above factors: poor data, variables considered important but not included for lack of data, and variables not included because they are not recognized as important.

Given the complex nature of the deforestation process and the obvious correlations among some of the independent variables, path analysis appeared to be an appropriate technique to apply in an attempt to articulate more fully the relationships among the independent variables and between the dependent and independent variables. In general, path analysis is a technique "for studying patterns of causation among a set of variables" (Dillon and Goldstein 1984, 431). Given the lack of data, figure 2 could not be tested in full. Instead, as presented in figures 4 and 5, simplified versions were analyzed, with and without forest area in 1970 as an independent variable. The statistical work was done using the Linear Structural Relations (LISREL) program, and path coefficients represent path correlations between variables. Only paths that are significant at the .05 level are illustrated. The model with forest area in 1970 (fig. 3) is significant at the .05 level, but the model without forest area in 1970 (fig. 4) is not statistically significant. The goodness of fit is greater than 0.96 for both models.

In the model which includes forest area in 1970 (fig. 3), the total path correlations with deforestation, in order of importance, are as follows: forest area in 1970 (0.653), distance (0.269), change in agricultural area (0.238), change in population (0.214), change in roads (0.077), and ACC in 1970 (–0.01). In figure 4, which does not include forest area in 1970, the total path correlations with deforestation are as follows: distance (0.437), change in agricultural area (0.376), ACC in 1970 (0.363), change in population (0.243), and change in roads (0.158).

Overall, the path analysis confirms the results of the panel analysis. The most important factors are forest area in 1970, distance from Manila, ACC in 1970, and the expansion of human settlements as represented by changes in agricultural area, roads, and population. The results indicate that the expansion of agriculture is more important than the increase in roads or population. In addition, the path analysis does not indicate that population increase is the driving force of deforestation.

Summary

When interpreting the results of this chapter, it is important to remember that the data cover a twenty-three-year period. It seems reasonable to admit the possibility that the process of deforestation itself could change over this length of time as lowland forests declined and remaining forests came to be found only in upland areas. In addition, Philippine society changed during this period and events external to the process of deforestation may have been important. Events worth mentioning include the declaration of martial law in 1972, the two oil price increases of 1972–73 and 1979, and large-scale fighting between the Philippine army and Muslims in Mindanao during the 1970s. In short, there is no a priori reason to expect the process of deforestation to be defined by the same parameters throughout this period.

The results of the PCAs and the multiple regression analyses agree with the framework set forth by this study in chapter 5. The presence of human settlement is negatively related to forest cover. Road density in 1957 and road and population density in 1970 and 1980 were the most important variables in explaining the variation in provincial forest cover. The results support the interpretation that the 1957 data were capturing the early stages of frontier settlement and the 1970 and 1980 data represent a more advanced stage of human settlement.

The panel analysis, which measured deforestation directly, revealed that the decrease in forest cover was positively related to ACC in 1970 and changes in agricultural area. These results were discussed above and support the analysis of chapter 5. That changes in population and population density were not important in explaining deforestation indicates that the factor of overwhelming importance was the size of the forest resource and its control by government and military officials.

It is interesting to contrast the results of the statistical analysis of this study with the review in chapter 3, regarding postwar discussions in the

Philippines on the causes of deforestation. That discussion has tended to blame either loggers or shifting cultivators for the decline of forest cover. The present research, on the other hand, has emphasized that these two agents of forest destruction cannot be considered separately; rather, they are part of the same process.

In addition, at a higher level of analysis, commercial logging and the movement of poor migrants to formerly forested areas is a reflection of a development process which has resulted in poverty for the mass of people and wealth for the few. The political economy of the forest resource in the Philippines is central to any analysis of deforestation.

8
Implications and Conclusions

Comparison with Quantitative Work on Tropical Deforestation

As the review in chapter 2 demonstrated, quantitative work on tropical deforestation is limited; in fact, there are only five studies at the global level (Allen and Barnes 1985; Grainger 1986; Palo et al. 1987; Rudel 1989; Scotti 1990), one at the regional level, for the Greater Caribbean (Lugo et al. 1981), and two at the subnational level, for northeast Thailand (Sungsuwan 1985; Panayotou and Sungsuwan 1989) and the Brazilian Amazon (Reis and Margulis 1990). Other than the present study, no national-level analysis of tropical deforestation on a quantitative basis has been conducted. While provincial-level forest-cover data based on LANDSAT images are available for all of Thailand for the years 1978 and 1982 (Bhumibhamon 1986) and India from 1972–75 to 1980–82 (National Remote Sensing Agency 1983), I am not aware of this data being used for a national study of deforestation.

A major difficulty with cross-sectional regression analyses dealing with nations is that the regression coefficients are virtually impossible to interpret, since they represent an average of all the nations in the sample. Sample sizes of the cross-national studies are as follows: Allen and Barnes (39 and 25), Grainger (30), Lugo et al. (29), Palo et al. (60), Rudel (36), Scotti (47). Thus, it is not clear how much, if anything, cross-national studies tell us about an individual country. The fear is that they may hide a great deal of significant variation among countries. Despite this limitation, the studies referred to above yielded similar results. The general findings of these studies are presented in table 30. Two different dependent variables were used; either a measure of actual deforestation (Def) or percentage forest cover (%FC). The "percentage deforested" (%Def) used by Reis and Margulis is simply the obverse of %FC.

Table 30.
Important variables in quantitative work on tropical deforestation

Unit (area)	Sample Size	R^2	Dependent variable	Independent variables
Municipality (Brazil)[a]	165	.84	%Def	+ population density + road density + crop area
Province (Thailand)[b]	64	.77	%FC	− population density − wood price + provincial income + distance from Bangkok
Nation (Caribbean)[c]	30	.65	%FC	− population density − energy use density
Nation (Global)[d]	60	.60	%FC	− population density
Nation (Global)[e]	47	.81	%FC	− population density
Nation (Global)[f]	36	.77	Def	+ population growth + availability of capital
Nation (Global)[g]	43	.52	Def	+ population increase + logging
Nation (Global)[h]	39	.11−.50	Def	+ population increase + increase in farmland + wood use

[a]Reis and Margulis (1990).
[b]Panayotou and Sungsuwan (1989). Four other variables were included in the final equation, but none of these were significant at the .05 level. As such, the final r^2 is overstated. They used pooled cross-section and time series data.
[c]Lugo et al. (1981).
[d]Palo et al. (1987).
[e]Scotti (1990).
[f]Rudel (1989).
[g]Grainger (1986).
[h]Allen and Barnes (1985).

Table 30 indicates that the r^2s obtained when percentage forest cover is the dependent variable are greater than when deforestation is the dependent variable. This general finding is not substantiated by the results of chapter 7, where the panel analysis yielded r^2s similar to the cross-section analyses. In short, it would appear that it is easier for the studies in table 30 to explain the distribution of percentage forest cover than deforestation itself. This raises the question of whether or not the two phenomena (distribution of forest cover and deforestation) are so different that they may not be directly comparable.

Of the eight research projects listed in table 30, only the works by Panayotou and Sungsuwan (1989) and Reis and Margulis (1990) are not cross-national studies and the dependent variable used in both was percentage forest cover. This means that, with the exception of my own panel analysis, no quantitative analysis of deforestation has been attempted for an individual tropical country. In addition, the dependent variable in the cross-sectional analyses in my study is absolute forest cover, as opposed to percentage forest cover in the other studies. As explained in chapter 7, absolute forest cover is more important as a draw for loggers and potential migrants than percentage forest cover.

All sources in table 30 agree that population growth and population density are important, if not the most important, independent variables explaining variation in deforestation and forest cover respectively. My own research indicates that population density was the most significant factor in the distribution of forest cover for 1970 but not for 1957 and 1980. In addition, the panel analysis for 1970 to 1980 did not show population growth to be an important factor in deforestation; however, the path analysis indicated that it does play a role, although not a major one.

Of particular interest is the relative lack of direct statistical evidence regarding the importance of the expansion of agricultural land in the process of deforestation. Lugo et al. (1981) state that the efficiency of the forest conversion process is inversely related to energy use per unit of land area, and they interpret this energy density variable as a substitute for the growth of agriculture; however, this is not direct evidence that expansion of agriculture is causing deforestation. Allen and Barnes (1985) did not find a statistically significant relationship between changes in forest cover and changes in arable land; however, based on the strong bivariate correlation between population growth and agricultural expansion, they concluded that changes in agricultural land are associated with deforestation. Allen and Barnes (1985) did determine a negative relationship between the percentage of land under plantation and change in forest area but term this a "surprising finding" because they cover such a small percentage of the area of the countries in their sample. They are not prepared to make the claim that expansion of plantations is causing deforestation. On the other hand, Reis and Margulis (1990) did find a positive correlation between deforestation and increases in crop area.

A possible explanation for the lack of significance of the expansion of agriculture, as pointed out by Allen and Barnes (1985), is the strong bivariate correlation between increases in population and changes in agricultural land. I would also add changes in road density to the above two variables. An additional consideration has to do with the available data. If expansion of agriculture is occurring in frontier/upland areas remote from population centers or government offices, this expansion may not be accurately reflected in official statistics. At the same time, it may be the case that population and road data are easier to gather; thus, there may be a built-in bias toward the underreporting of agricultural land vis-à-vis population and road networks.

Lastly, even assuming that an error-free agricultural census with 100% enumeration is available, it is still most likely that expansion of agricultural land would not be accurately captured, simply because former agricultural lands, which are now abandoned, will not be recorded as agricultural land, i.e., land actually farmed. The extent of this phenomenon will depend upon the quality of land, population density, and myriad other factors. Since abandonment of land has been cited by virtually every observer of tropical deforestation, it is probable that agricultural censuses every ten years do not reflect this activity. This discussion raises the very difficult question of how abandoned land is defined and measured. A brief review of what few data exist on these areas in the Philippines will demonstrate the empirical difficulties involved.

The NEC (1959) reported that brushland (dominated by shrubs and brush) and open land (dominated by grass) covered 20,772 and 34,029 (total = 54,801) km^2 respectively in 1957. The results of the second national forest inventory (*PFS* 1973) indicate that open land was equal to 7,706 km^2 and reproduction brush was equal to 41,579 km^2 (total = 49,285). The FDC (1985) ascertained that open land was equivalent to 50,300 km^2 in 1980. The SSC (1988), using SPOT imagery, determined that grassland was equal to 18,129 km^2 and "cultivation mixed with brushland and grassland" was equal to 101,143 km^2 (total = 119,272).

The results of these four national surveys indicate that they do not share common definitions or categories of land use. If correct, open land was 34,029 km^2 in 1957; 7,706 in 1969; 50,300 in 1980; and 18,129 in 1988. It is difficult to imagine a process which could produce these numbers. The variation among the surveys is so great that it is not even possible to determine whether open land is increasing or decreasing at the national level. At the provincial level, the discrepancies among the surveys are even greater. In short, the data do not permit us to make even general statements about what is obviously a major land-use category. This confirms earlier statements about the lack of detailed studies concerning conversion of forests to agriculture and whether the agriculture which follows is permanent or shifting.

This point has important implications for future research on tropical deforestation. Our lack of knowledge at the micro and macro levels of the land-use category in between agriculture and forestry (abandoned/open land) is a major stumbling block to effectively understanding the process of deforestation. We simply do not have accurate information on the number of migrant farms or how often they shift. Nor do we know whether grasslands as opposed to forests are being converted to farmland (see Dove 1983 for a case study in Indonesia where grasslands are used for long-term agriculture). Our understanding of tropical deforestation will necessarily be incomplete until this important gap in our knowledge is filled. The statistical insignificance of agriculture as an explanatory variable of forest cover or deforestation in the studies listed in table 30 may be explained by the above considerations.

On the other hand, not only is it likely that road and population data are easier to gather than agricultural data, but roads and population are not as easily abandoned as agricultural land. Even if frontier agriculture moves further into the forest, roads remain and continue to be used. Population will continue to grow from natural increase and in-migration. In short, roads and populations will be systematically overstated relative to land in agriculture. If correct, new methods of data collection may be necessary to understand the farmland/forest/open land interaction.

By way of contrast, the panel analysis of this study has found that changes in agricultural land from 1970 to 1980 were significant and positively related to deforestation. This result was confirmed by the path analysis. In addition, a regression of the change in agricultural land from 1970 to 1980 as a function of logging in 1970 yields an r^2 of 0.15 and the result is significant at the .05 level. This indicates that the expansion of agriculture is, to a certain extent, a function of logging in a preceding period.

Of the eight studies listed in table 30, Allen and Barnes (1985), Lugo et al. (1981), Palo et al. (1987), and Scotti (1990) did not include a variable for roads. Grainger (1986) argues that accessibility to forests should be important in explaining forest cover/deforestation and mentions proximity to rivers and road infrastructure as relevant considerations. He did not include data on roads, but he used area logged previously as a proxy for logging roads and this variable was significantly correlated with deforestation. The variable for river density was poorly correlated with deforestation.

Panayotou and Sungsuwan (1989) and Reis and Margulis (1990) are the only authors to explicitly include a variable for road density. The road density variable of Panayotou and Sungsuwan (1989) is for village roads only and measures roads completed by the Office of Accelerated Rural Development during the period of their study, 1973–82. As such, this variable does not include nonvillage roads, roads in place before 1973, or logging roads. Their results indicate that road density was not significant at the .05 level. Reis and Margulis (1990), on the other hand, found a positive correlation between deforestation and road density, and Rudel (1989) determined that large-scale capital-intensive projects such as the trans-Amazon highway result in deforestation.

My own study has determined that road density was an important explanatory variable of forest cover in 1957 and 1980. In addition, the increase in kilometers of roads between 1970 and 1980 was positively related to forest area in 1970. Lastly, in a simple regression, deforestation was positively related to the increase in roads between 1970 and 1980. In other words, I have found roads to be a significant explanatory variable of forest cover and deforestation. Furthermore, the apparent declining importance of road density indicates that roads are more important at the beginning of the deforestation process or when an area is first opened up. The inclusion of a road variable is a major difference between most of the studies listed in table 30 and the present research.

The panel analysis of this study indicated that deforestation in the 1970–80 period was positively related to forest area in 1970, ACC in 1970, distance from Manila, and changes in agricultural area. Of these four variables, both Grainger (1986) and Allen and Barnes (1985) have concluded that forest area is important, Grainger in his correlation and multiple regression analysis, and Allen and Barnes by including percentage forest area in their regression analysis. However, my own results do not agree with those of Allen and Barnes (1986). They found that countries with a low percentage of forest cover in 1968 were "more likely to lose [absolute] forest area" during the 1968–78 period, a conclusion I find very difficult to reconcile at the theoretical and empirical levels, and in conflict with results showing deforestation to be positively related to absolute and percentage forest cover. In addition, both studies concluded that logging was important: Grainger (1986) by including logging as a variable in his regression analysis, and Allen and Barnes (1985) through their inclusion of a wood-use variable which includes wood exports and fuelwood consumption.

It is important to note that Grainger (1986), Allen and Barnes (1985), and Rudel (1989) are the only studies which have deforestation as the dependent variable. In the cross-sectional studies, neither forest area nor logging could be included as an independent variable, since forest area is the dependent variable and logging at any moment in time is a function of forest cover rather than vice versa. In this regard, cross-sectional studies of forest cover are necessarily limited.

The distance variable in my panel analysis has no counterpart in the studies listed above, with the exception of Panayotou and Sungsuwan (1989). They argue that forest cover in northeast Thailand is positively related to distance from Bangkok for two reasons: first, the major market for hardwoods is in Bangkok and higher transportation costs from the more remote areas reduce their profitability; second, as distance from Bangkok increases, the demand for land to grow cash crops decreases (presumably because Bangkok is the ultimate market for most of these crops). On the other hand, my own conclusion is that deforestation is positively related to distance from Manila. The two results do not necessarily contradict each other; rather, they demonstrate the difference between having forest cover or deforestation as the dependent variable.

In the Philippines, forest cover is positively related to distance from Manila, as in Thailand, but it is incorrect to draw any inference regarding deforestation on this basis alone. That deforestation increases as one moves away from Manila is explained by two factors: increased logging away from Manila and less control of the forest removal process. The negative correlation between forest cover and population density obtained in the cross-sectional analyses obscures the fact that deforestation is actually occurring more rapidly in areas with low population densities. The cross-sectional analysis captures the end result of a historical process, but it cannot capture the dynamic process itself.

In the case of the distance variable, forest cover is positively correlated to distance in both Thailand and the Philippines. On the basis of this relationship, Panayotou and Sungsuwan (1989) concluded that deforestation is greater the closer one is to Bangkok; however, their evidence does not support this conclusion. The difficulty arises from assuming that a high percentage of forest cover means that deforestation is not occurring or is occurring at a slow rate. This assumption is not correct. The crux of the matter is that Panayotou and Sungsuwan (1989) have confused forest cover and deforestation. Their stated concern is "to assess the relative importance of the different determinants of the forest cover (or inversely, causes of deforestation)" (p. 21). As the present research has indicated, deforestation is not the inverse of forest cover.

The differences between using forest cover and deforestation as the dependent variable are dramatic in terms of the level of explanation, independent variables included in the analysis, and interpretation of the results. A major conclusion of this study is that studies of forest cover are of limited value to understanding the process of deforestation.

An additional consideration in the interpretation of the distance variable is that the physical nature of each country is vastly different. The Philippines is an archipelago and Thailand is basically a compact country. In the Philippines, log exports move directly from the producing island to the overseas buyer; in particular, they do not have to go through Manila. In Thailand, on the other hand, legal log exports have had to go through Bangkok and transport is primarily by road. Thus, the notion of log transport cost as a function of distance from the capital city makes sense for Thailand but not for the Philippines. However, even this must be qualified by the notion that, since economic rents in forestry have been great, marginal changes in transportation costs may not be a significant factor in logging operations. The recent logging by Thai firms in northeast and eastern Burma indicates that these costs are not an inhibiting factor for large-scale and politically well-connected Thai loggers (*Asiaweek* 1989).

Another difference between Thailand and the Philippines regards the extent of cash crops. While the spread of cash crops has been reliably cited as a factor in deforestation in Thailand, this does not appear to be the case in the Philippines. If removal of forest cover is followed by subsistence agriculture, then transport to Manila is not a relevant consideration. This is not intended as criticism of Panayotou and Sungsuwan (1989); rather, it underlines the fact that the process of deforestation is not the same for every nation. Obvious differences of potential significance include the physical nature of the country (mainland versus insular), the presence or absence of commercial logging (dipterocarp versus nondipterocarp forest), types of agriculture practiced, and climate (presence or absence of a dry season).

Detailed case studies of tropical deforestation are needed to capture the uniqueness of each nation's deforestation process. The lessons to be learned from a study of loss of forest cover in the Philippines are most likely more applicable to Indonesia than to Brazil. Generalizations based on cross-national studies may be of little relevance for formulating specific programs to control deforestation. Also, studies using forest cover as the dependent variable may be of limited value to understanding the dynamic process of deforestation.

Of all the studies discussed above, only Panayotou and Sungsuwan (1989) include a price variable by incorporating prices for kerosene (a wood-fuel substitute), crops, and wood (logs). They argue that, after population density, the price of wood is the "second most important determinant of forest cover" (p. 22). According to their economic model, higher log prices lead to greater profits which lead to greater logging which leads to deforestation. They note that logging is a source of deforestation but not a cause, because the real culprit is the insecurity of logging concessions, which encourages a cut-and-run attitude on the part of the loggers and discourages proper management, replanting, and protection. In short, destructive logging in the tropics is a rational response to the institutional setting that loggers find themselves in. I would also add that the institutional setting is to a large degree determined by the loggers themselves through their contacts with influential politicians, and the distinction between government regulators and regulated loggers is often a thin one.

While I agree with Panayotou and Sungsuwan (1989) that logging is an important factor in tropical deforestation, the role of log prices in this process, at least in the Philippines, needs to be qualified. It should be obvious that logging occurs because it is profitable; in fact, as chapter 3 pointed out, the economic rents accruing to the forestry sector in the Philippines have been enormous. These rents exist because of deliberate government policy.

In a situation where large economic rents exist, marginal changes in price most likely have little effect on the economic activity itself. Whether a logger's economic rent is 1,000 or 1,100 pesos per cubic meter is not so great a difference as to significantly affect logging behavior. The insecurity of tenure that Panayotou and Sungsuwan (1989) talk about ensures rapid logging, since this is the only way the economic rent can be captured. Given that Segura-de los Angeles et al. (1988) have reported profit margins of 99–182% for logging in the Philippines, short-term changes in the price of logs will most likely not affect logging either way.

In an environment where government policy guarantees large rents, where mismanagement is not punished, and where large-scale collusion exists among government officials, the military, and loggers, price per se is a secondary concern. The real issue is the political economy of logging in the tropics, especially Southeast Asia. To state that logging occurs because it is profitable is obvious but does not help us to understand why it is so profitable, why only a small group of people benefit from it, why massive corruption is the norm, and why virtually nothing has been done in the Philippines (or Thailand) to stop it. While it cannot be proven empirically, it seems clear that government control of logging concessions and tolerance, if not encouragement, of illegal logging and smuggling have been more influential in determining the rate of logging than the price of logs. The economic rents documented by Repetto (1988) support this line of reasoning. Palo et al. (1987) deliberately did not include prices in their statistical analysis of tropical deforestation because they feel that profits, prices, and costs have only an indirect, minor role in deforestation. I agree with this assessment.

Even though the estimates of Repetto (1988) regarding economic rents in the forestry sector are reasonable, his conclusion that governments have adopted misguided policies does not necessarily follow. It is important to reiterate that the process of deforestation, at least in those countries where the forests have a commercial value like the Philippines, Indonesia, Malaysia, and Thailand, has greatly benefited a relatively small group of people. Although the process may appear to be chaotic, enough control has been exercised from the top to ensure a continuous flow of financial rewards. The process has been more manipulated than misguided. For those who have benefited from the process, it has served their purposes well.

The point is that governments in many countries do not represent the public; rather, they represent elite interests. In such countries, the major interest of the elites has been to enrich themselves, and the Philippines is no exception (Korten 1986). In the Philippines, this is called "private use of public office" and is culturally acceptable behavior. Steinberg (1967) provides an example of such predatory behavior on the part of the Filipino elite that is an apt metaphor for their general behavior in the postwar period. In early 1945, the Philippine Congress convened for the first time since the Japanese invasion in 1941. World War II had devastated the country and Manila lay in ruins. The Philippine treasury was almost bankrupt. And yet the first action taken by the congressmen (many of whom had collaborated with the Japanese) was to vote themselves three and a half years' back pay for the war years. While this may appear to be an extreme example, it does demonstrate the "rapacious" nature of the Philippine elite (Oshima 1987) and helps explain why Bello (1988) has termed Philippine government in the postwar period a form of "institutionalized looting."

The role of domestic log prices in deforestation must be qualified by an additional consideration: the importance of foreign exchange. It may be that the desire of loggers and those allied with them is to earn not pesos but dollars or yen. If this is the case, domestic log prices would not be an accurate reflection of the economic profitability or desirability of logging. Unfortunately, data on this point are almost nonexistent. However, in addition to the cases of dollar salting discussed in chapter 3, several factors make it highly likely that this was an important consideration in the Philippines: foreign exchange has been in short supply in the Philippines since the late 1940s (a large black market for foreign exchange is evidence of this) and the peso has steadily declined in value relative to the dollar/yen in the postwar period. In short, part of the driving force of commercial logging may be the foreign exchange that can be derived from selling logs overseas rather than the domestic price.

The remaining variable that Panayotou and Sungsuwan (1989) found to be statistically significant was provincial income, which was positively related to forest cover. They suggest that higher provincial income lowers demand for fuelwood and reduces the demand for conversion of forestland to agriculture. Since my study did not include a variable for provincial income, a direct comparison is not possible. However, several comments are in order.

First, although Allen and Barnes (1985) agree with the hypothesis of Panayotou and Sungsuwan (1989) that higher per capita income should lead to the substitution of commercial fuels for fuelwood, as Anderson (1986) points out, this effect appears to be weak in low-income countries. In the case of the Philippines, where absolute and relative poverty have been increasing for decades, it is doubtful that this relationship would hold. Interestingly, Allen and Barnes (1985) did not find per capita GNP to be a statistically significant variable. Palo et al. (1987) found GNP per capita to be statistically significant at the .10 level but it was negatively related to forest cover.

Second, as discussed in chapter 6, it is doubtful that fuelwood collection per se has led to deforestation. While commercial charcoal making has no doubt led directly to deforestation in some instances, fuelwood gathering, at least in the Philippines, cannot be considered to be a major cause of loss of forest cover.

Third, while Panayotou and Sungsuwan (1989) assume that higher provincial income will reduce incentives to convert forest to agricultural land, Hyde and Hamilton (1988) argue that the income effect is uncertain because higher incomes may lead to more consumption of fuelwood or a switch to other fuels.

In short, on a theoretical level, any hypothesized relationship between deforestation and economic development must be examined more carefully before any conclusions can be reached. It may be that in some countries this relationship will be positive and in others negative. The empirical evidence is contradictory. The relationship may not be the same for all societies, particularly if the distribution of income becomes more highly skewed while per capita income increases. This highlights once again the importance of detailed national-level studies.

I have pointed out that one of the major difficulties in discussing tropical deforestation is the uncertainty regarding rates and extent of deforestation and the ambiguity surrounding the terms "deforestation" and "degradation."

Unfortunately, this conceptual confusion exists in the few quantitative studies that have been done. The eight studies listed in table 30 and my own study have used a total of three dependent variables; absolute forest cover, percentage forest cover, and loss of forest cover. Each study has used a different set of independent variables. Most work has been either cross-sectional or cross-national or both. Only three studies (the Philippines, Brazilian Amazon, and northeast Thailand) are at the national or subnational level. To a certain extent, the quantitative work conducted so far on tropical deforestation is not directly comparable.

One of the obvious conclusions of the present research is that the almost complete lack of rigorous studies of tropical deforestation at the national level or below seriously hampers our ability to make any general statements regarding tropical deforestation or, indeed, to conceptualize the process adequately. That authors who have studied only forest cover claim to have studied deforestation is a case in point.

A major distinction must be drawn between using loss of forest cover or actual forest cover as the dependent variable. The distinction is twofold. First, at the empirical level, r^2s with forest cover as the dependent variable are, in general, much higher than for deforestation. As such, the limited quantitative work suggests that these two phenomena are not the same. At the very least, the choice of dependent variable will effect the independent variables that can be included in the analysis. If forest cover is the dependent variable, forest cover and logging cannot be used as independent variables. In addition, if forest cover is the dependent variable, the independent variables will, of necessity, be density variables. On the other hand, if deforestation is the dependent variable, then the independent variables will represent absolute changes.

Second, at the conceptual level, there is reason to assume that the forces which determine forest cover are not necessarily the same as those determining rates of loss of forest cover. The amount of forest cover remaining today is the end result of a process which may have started hundreds or thousands of years ago. That forest cover today is negatively related to population density does not mean that population density today is even an important explanatory variable of deforestation. Indeed, logically speaking, this is not likely. Current deforestation occurs where forests presently exist, and where forests exist now, population densities are generally low. However, this does not necessarily mean that high population densities cause deforestation. The fact that North America, Europe, and Japan have all experienced increasing forest area in the postwar era should caution us not to assume that increased population must cause deforestation. Deforestation has been rapid throughout Southeast Asia, yet population densities are vastly different among these countries. In fact, it would be difficult to argue that Malaysia faces any serious population pressure at all.

Considering the above, it is inappropriate to make statements regarding the process of deforestation based on an analysis of forest-cover data alone. An analysis of forest-cover data for several different years may be appropriate, but the very different results of the panel analysis as opposed to the cross-sectional analyses in this study indicate that forest cover and loss of that forest cover are the result of different activities.

Another way to express the notion that forest cover and deforestation represent two entirely different phenomena utilizes the concept of stocks and flows. Forest area is a stock; it represents the sum total of past influences on the forest resource. Deforestation is a flow; it represents the recent marginal change (negative) of the stock. The present stock of forest cover is the result of two forces; the sum total of marginal negative (deforestation) and positive (regeneration) changes.

Socioeconomic density variables such as population will almost always be negatively related to forest cover, since they represent human settlements; and where there are human settlements, by definition, there are fewer forests. However, since density variables represent the sum total of the history of human occupancy, they tell us little about present-day deforestation. As an extreme example, population density and forest cover in parts of China are the result of four thousand years of history. Since the subject matter is ongoing deforestation, our concern should be with marginal changes of forest cover and the relevant socioeconomic variables.

This study raises questions about the value of cross-national regression analyses in understanding deforestation in a particular country. Regression coefficients derived from such research cannot be used to design policy instruments at the national level and cannot substitute for detailed national studies. They are unable to capture the unique aspects of deforestation within individual countries. In our limited understanding of the process of deforestation, the most critical gap is the lack of detailed national and subnational studies.

Theory

The empirical results of chapter 7 and the extended discussion in chapter 3 regarding Philippine forestry permit comparisons to be made with the theoretical work on tropical deforestation reviewed in chapter 2. Like the quantitative work, theoretical work on tropical deforestation is rather limited; however, all of the works reviewed in chapter 2 are of direct relevance to this study.

The work of Guppy (1984) provides an appropriate framework in which to view Philippine deforestation. As discussed in chapter 2, Guppy (1984) connects global deforestation with four separate but related phenomena that bear repeating here:

Factor 1. Rapid population growth and land hunger, which are the ostensible causes of deforestation.

Factor 2. Inequitable access to resources (forests and agricultural land).

Factor 3. The political motivations of the local political and business leaders, which are primarily to enrich themselves and maintain power.

Factor 4. The active support these leaders have received from Western governments, which has enabled them to stay in power.

The scenario presented by Guppy (1984) is remarkably similar to the process of deforestation that has occurred in the Philippines. The Philippine government, particularly after President Marcos declared martial law in 1972, received massive foreign funding in the 1970s and early 1980s, with foreign debt increasing from U.S. $3.4 billion in 1975 to US $19.8 billion in 1985 (NEDA 1987). In addition, the martial law regime had the strong support of the American government (Bonner 1987). Rates of deforestation in the Philippines were at their highest in the 1970s.

The desire of the Philippine elite to enrich themselves through controlling access to the primary forest resource has been discussed throughout this study. I would add, however, that this preferential access has been accompanied by the deliberate manipulation of the statistical reporting system and of the agencies ostensibly in charge of the forestry sector. Like Guppy (1984), I would argue that the deforestation which occurred in the Philippines between 1970 and 1980 was a function not of population pressure but of commercial logging and lack of development in the lowlands, which forced people to engage in subsistence agriculture in areas formerly occupied by primary forest. As the DENR et al. (1987) concluded, natural resource use in the Philippines has been characterized by the same sort of inequities which characterize Philippine society in general: elite control and limited access for the poor.

In short, the model proposed by Guppy (1984) fits the Philippine case well. I would also note the commercial importance of the Philippine forests and the importance of overseas markets for Philippine wood products. Guppy's model may be most relevant to those countries where commercial logging has been important: Western Africa and Southeast Asia. Overall, Guppy's emphasis on the political economy of forest destruction is appropriate for the Philippine situation.

The regional political ecology approach of Blaikie and Brookfield (1987) has as its main concern the land manager (the person who actually makes the immediate decisions regarding land use) and the political/economic/ecological context in which these decisions are made. The central question is why land saving/improving investments are not generally being made. In the

Philippines, the two main users of forestlands are loggers and agriculturists, and their reasons for not undertaking investment in the land are different.

Loggers have not, in general, invested in proper forest management or forest protection for two reasons. First, since their attitude was to "cut and run," they were solely interested in maximum returns in as short a period as possible. Second, there was no incentive for them to properly manage the forest resource, since negative sanctions were rarely, if ever, applied. On the other hand, agriculturists have had little incentive to invest in land improvement, since in most cases they did not have a proper land title to what was technically government land. In addition, negative externalities as a result of excessive erosion were borne by third parties downstream.

A consideration which may have been operating in formerly forested lands is what I will call the "demonstration effect" of unrestricted logging. Most likely, agriculturists and people in the uplands were acutely aware that logging was widespread and conducted in a destructive manner. Under these circumstances, it may have been quite easy for uplanders, particularly migrants, to treat the forest resource in a manner equivalent to that of the rich and well-connected. While this point cannot be proven, my study suggests that the rules of the game regarding forest utilization were established by the logging concessionaires, and the poor and disenfranchised simply followed suit.

An advantage of the work of Guppy (1984) and Blaikie and Brookfield (1987) is that the class nature of society and differential access to the sources of societal power are explicitly recognized. This advantage is also shared by the works reviewed on disaster theory (Hewitt 1983; Susman et al. 1983; Watts 1983). The creation and maintenance of marginalized people in Third World societies has led to the migration of the poor to upland areas, many of them previously forested. Once again, the importance of elite control and lack of access to resources on the part of the poor are stressed. The vulnerability of the poor in marginal environments is a function of the entire social system (elite control of resources and the increasing poverty of the majority of citizens). Deforestation is one of the end results of this increasing vulnerability and, at the same time, it helps to make the poor even more vulnerable. The value of the works on disaster theory is that they emphasize the interaction between society and the environment and place at the center of analysis the hardships suffered by an increasingly large group of people.

Grainger (1986, 1987) has written the most extensively about modeling tropical deforestation. He argues that while logging and the expansion of agriculture are the types of forest exploitation that result in deforestation, the "primary causative factors" are population growth, economic development (increase in per capita income), and accessibility of the forest resource. The driving forces in his model are increases in population and GNP, which increase demand for food and wood products and at the same time make possible investments in agriculture, which can increase agricultural yields and thus slow down deforestation. Grainger's attempt at modeling this process can be viewed as an effort to describe deforestation more precisely and to articulate the linkages among the economic, forest, and agriculture sectors. He does not discuss the political economy of deforestation, which I have determined to be crucial for understanding deforestation in the Philippines.

Both Walker (1985, 1987) and Panayotou and Sungsuwan (1989) discuss the institutional rules under which logging concessions operate. Both agree that length and security of tenure are the key issues facing loggers and that short-term leases and insecurity of tenure are the main reasons for excessive and destructive logging. In short, deforestation through logging results from a failure on the part of governments to "get the prices right." While this is correct in a theoretical, economic modeling sense, such a conceptualization ignores some important factors. Neither author defines security of tenure in a meaningful manner. If security is equated with length of tenure, then presumably a 100- or 200-year concession agreement would represent long-term security and provide sufficient incentive to husband the forest resource for future harvests. Unfortunately, I cannot accept this as an adequate representation of the notion of security; as Segura-de los Angeles et al. (1988) have noted, increasing license tenure "does not significantly increase the firm's profitability," because of the effect of discounting.

Under present-day circumstances in Southeast Asia, commercial forests are a fleeting resource; their commercial timber value can be captured only by logging. Security does not lie in concession rights but in contacts with politicians, bribes to government officials, and payoffs to military personnel (Hackenberg and Hackenberg 1971). Accordingly, logging takes place in a fluid arena and loggers, at least in the Philippines, attach a high-risk premium to future benefits (Segura-de los Angeles 1986). In all of the major countries of Southeast Asia (Burma, Indonesia, Malaysia, Philippines, Thailand), corruption is widespread and democratic control is limited. When access to the forest is permitted, the only rational economic course of action is to cut as quickly as possible, because this access may be cut off in the future as a result of a change in government or because another party has offered a larger bribe. Since neither of the authors above can define security of tenure in the context of Southeast Asian society today, their economic models are of limited value for understanding excessive logging in tropical countries.

Economic models ignore the fact that often it is loggers who decide upon the institutional context in which they operate. They (or their allies) are also called upon to enforce the rules of the game. Economic modeling which ignores these basic facts of political economy simply cannot capture the major forces behind destructive logging, at least in Southeast Asia. Under present circumstances, money in the bank (especially an overseas bank) represents more security than possible future returns from a properly managed forest resource. The arguments of Walker (1985, 1987) and Panayotou and Sungsuwan (1989) assume that proper forest management is possible in Southeast Asia under present-day circumstances, an argument I do not accept.

Overall, the theoretical work that best agrees with this study is that which emphasizes the political economy of resource use in tropical Third World countries. At the same time, this does not necessarily mean that there is one general, unified theory of tropical deforestation applicable to all countries. This is particularly the case when a longer-term perspective is considered. Deforestation in India has been going on for several hundred years, whereas deforestation in the Brazilian Amazon has only really occurred in the past two decades. There is no reason why one theory should be able to explain both of these cases adequately. The political economy element of deforestation is important in Southeast Asia because the forests are commercially valuable; however, this may not be an important consideration in other areas. This issue can be resolved only when more detailed national and subnational studies of deforestation become available.

Population Pressure and Deforestation

This study raises serious questions regarding the way that the expansion of human settlements in the tropics and the resulting deforestation are conceptualized. The results of the panel analysis showed that increases in population were not important in explaining the variation in deforestation across provinces. The path analysis indicated that changes in population contributed to deforestation but that they were overwhelmed by the influence of forest area in 1970, distance from Manila, logging in 1970, and changes in agricultural land.

It is clear from the panel analysis that population pressure itself does not result in deforestation, at least in the Philippines. At the same time, there is ambiguity in the term itself. A basic question is how population pressure is to be measured. Several standards are available: total population, population density, physiological density (number of people per unit area of arable land), percentage increase in population, and absolute increase in population. An additional measure could be in-migration, since migrants, as opposed to local inhabitants, may be moving to frontier areas. Lastly, it could be argued that whatever measure of population pressure is adopted, it should be lagged to take into account that children are less likely to create new farms than offspring who are old enough to set up their own households. Thus, in addition to the several ways to measure the forest cover/deforestation variable, there are at least seven means available to measure population pressure. It is not clear which measure is the most appropriate for studies of deforestation and, remarkably, there is little discussion in the literature on this topic.

It is important to keep in mind that the two measures of population pressure most commonly used in quantitative work on tropical deforestation (absolute population growth and population density) are not necessarily consistent. For instance, an area experiencing rapid population growth could have either a low population density (for example, a forested area) or a high population density (for example, an urban area). Deforestation occurs where population densities are low since, logically speaking, large-scale forests preclude large-scale human settlements. Thus, an area of low population density and high population growth which is experiencing deforestation does not demonstrate that population pressure is a problem within the area; rather, it may demonstrate that population pressure or social conditions outside the area are causing people to migrate. A good example of this would be what is occurring in the Brazilian Amazon (Millikan 1988).

The schema representing the deforestation process in the Philippines (fig. 2) has two main elements. First, the primary forest is converted to secondary forest through logging and, second, human settlements, primarily composed of poor farmers, then encroach upon the secondary forest. Implicit in this scenario is elite control of access to the primary forest and a development process which has created millions of poor people who have no place to go other than the forest. Interestingly, the large-scale migration of poor people to the formerly forested areas has occurred with the blessing of those in power.

Many observers agree that large-scale migrations in the postwar Philippines (particularly to Mindanao) have acted as a safety valve (Anderson 1982; Molloy 1983; Wernstedt and Simkins 1965). Political unrest and peasant discontent have been widespread in the postwar Philippines (Krinks 1983) and migration has been seen as a way to dissipate this unrest and lessen the obvious burden of an increasing population on a backward agricultural sector. These population movements thus made it easier for the elite to continue appropriating the primary forest resource by lessening calls for any fundamental changes in the socioeconomic system (Krinks 1974). Migration and deforestation have not in any way challenged elite control of the socioeconomic system.

In short, while population numbers and density are rapidly increasing in the Philippines, it is more important to understand the context in which this is occurring. Deforestation in the Philippines is a reflection of an entire socioeconomic system and the international context in which it is developing. A simple statement that increasing population and deforestation are correlated does not further our understanding of the process of deforestation.

Another consideration in the population pressure debate has to do with cross-national comparisons. Population pressure as a cause of deforestation does not apply to a large number of areas, particularly North America, Western Europe, and Japan (Ives and Pitt 1988; Totman 1989). In other words, it is not inevitable that increases in population must lead to deforestation. In addition, there is such a wide range of population densities in the tropical

Third World that it is impossible to draw any firm conclusions regarding the effect of population pressure on deforestation. For example, in 1990, population per square kilometer for five selected countries was as follows: Brazil (18), Indonesia (100), Malaysia (54), Thailand (108), Philippines (220) (Population Reference Bureau 1990). Absolute rates of deforestation are higher in Indonesia, Malaysia, and Thailand than in the Philippines; yet, Thailand has only 50% and Malaysia 25% of the population density of the Philippines. Brazil, which most observers would agree has the most extensive ongoing deforestation, has a population density which is only 8% that of the Philippines. These examples do not exhaust the possibilities; many countries in South America and Africa with population densities less than Malaysia are experiencing rapid deforestation.

While population obviously plays a role in tropical deforestation (somebody is cutting down the trees), a discussion which ignores the context in which population growth is occurring is incomplete (see Ives and Pitt 1988 for a discussion of this issue in the context of Nepal). The major timber-producing countries of Southeast Asia have had rapid population growth, but in addition they have experienced elite control of government and deliberate government policies which have encouraged large-scale commercial forest exploitation.

Another reason for doubting that population pressure is the main cause of tropical deforestation has to do with the rapidity of deforestation in the postwar period and why it occurs at certain times. The clearest example of this phenomenon today is the province of Rondonia in Brazil. The rates of deforestation documented by Fearnside (1986) and Millikan (1988) are impossible to reconcile with demographic factors. While in-migration has been rapid and involves large numbers of people, the resulting deforestation is not a function of the population density or population growth rate of the Brazilian population (Millikan 1988). Rather, it involves millions of extremely poor people in other parts of Brazil who are desperate to provide their families with at least a subsistence living. Road building and migration meet the needs of a Brazilian elite anxious to avoid or lessen tensions in the impoverished northeast and cities of southern Brazil. The colonization of the Amazon and the resulting deforestation is a substitute for reform within larger Brazilian society (Millikan 1988).

The rapid increase in deforestation in Indonesia after 1966 was not the result of population pressure. Rather, after President Sukarno's ouster, it reflected President Suharto's views on development in general and utilization of forest resources in particular. Rapid deforestation in Malaysia started around 1960 after the Communist insurgency was put down, not because of exploding population pressure. Even more telling is the recent history of forest destruction in Malaysian Borneo. Deforestation is now rapid in Sarawak and Sabah, provinces with population densities of 10.4 and 10 per square kilometer respectively in 1980. In the Philippines, rapid deforestation in Mindanao after the 1950s followed large-scale road building and the Philippines as a whole experienced high rates of deforestation in the 1970s, most likely after the declaration of martial law in 1972. In short, population pressure cannot explain the rapid initiation of deforestation at various places around the globe in the postwar period.

In Southeast Asia, there appears to be a clear relationship between logging and deforestation. In all cases, the decision to engage in large-scale logging has had government approval. It is difficult, if not impossible, in the examples just given, to see what role population pressure played. In fact, a simple correlation between population pressure and deforestation cannot explain why rapid deforestation is initiated in one area at a particular time. If the population pressure argument is correct, gradually expanding population would lead to a gradually declining forest cover. This has not happened in Southeast Asia. Just as areas of rapid deforestation could be found in the nineteenth-century Philippines as a result of the expansion of commercial agriculture, similar areas can be found in postwar Southeast Asia as a result of the expansion of commercial logging followed by subsistence agriculture.

Remote Sensing and Monitoring of Forest Cover

Remote sensing by satellite is increasingly being suggested as a way of monitoring the rapid changes occurring in the extent of tropical forests (Grainger 1983; Myers 1988b; Woodwell et al. 1983). Remote sensing offers the ability to evaluate forest and other vegetative cover over large areas on a repetitive basis and at a low cost per unit area covered. In an attempt to clarify the contribution that remote sensing has made or could make to analyzing vegetative cover, this section will briefly review the Philippine experience in this regard. Table 31 presents the five forest surveys which have been conducted in the Philippines using remotely sensed data.

Of the five studies in table 31, one was a computer analysis of digital data and four were visual interpretations of LANDSAT/SPOT images. The 1973 study was primarily conducted by the General Electric Co. and all com-

Table 31.
Forest surveys using remotely sensed data

Year	% Forest cover	Source of data	Method of interpretation	Source
1973	38.0%	LANDSAT	Computer	Lachowski et al. (1979)
1974	29.8%	LANDSAT	Visual*	Bruce (1977)
1976	30.0%	LANDSAT	Visual*	Bonita and Revilla (1977)
1980	25.9%	LANDSAT	Visual*	FDC (1985)
1987	23.7%	SPOT	Visual**	SSC (1988)

*Visual interpretation of black and white photomosaic images.
**Visual interpretation of color SPOT images.

puter work was done in the United States. The 1987 study was by the SSC and all of the interpretative work was done in Sweden.

It is interesting that the 1973 forest inventory by the General Electric Co. and the Department of Natural Resources indicated that 38% of the Philippines was under forest cover. This result is inconsistent with later forest inventories. The most probable reason for the very high estimate of the 1973 study is that the two land-cover categories of "forest partial closure" and "forest obscured by clouds" represent almost 50% of total forest cover. In retrospect, it is obvious that this inventory miscalculated forest cover by approximately 25%, since actual forest cover was 30% of national land area. In other words, forest cover was overstated by approximately 24,000 km^2.

The General Electric/Department of Natural Resources survey was initiated in 1976 but used images from 1972–74. I have taken 1973 as the midyear date. However, the NRMC in the late 1970s claimed that the images were from 1976. While a difference of three years may appear to be a benign mistake on the part of the government, it makes a major difference during a period of rapid deforestation in calculating rates of change in forest cover. This mistake is still current, as Myers (1988a), one of the leading experts on tropical deforestation and familiar with the Philippine situation, notes that the General Electric/Department of Natural Resources survey represents the forest cover of 1976.

A constructive feature of this first computer-assisted inventory was that it indicated that forest cover had dropped considerably from previous forest inventories. However, this must be weighed against the fact that forest cover was overstated. Lastly, forestry officials should have been aware that rapid deforestation was occurring, considering that the second national forest inventory had already been printed in the *PFS* (1973).

Some of the difficulties with the SSC survey using SPOT data from 1987 were discussed in chapters 3 and 4. While the overall forest-cover figure appears to be reasonable, it is very difficult to reconcile specific forest-cover categories at the national level and forest area at the provincial area with those of the P-GFI. For reasons already stated, I find the P-GFI data to be more reliable.

The three inventories conducted visually using LANDSAT imagery from 1974 to 1980 were all done by Filipino experts (Dr. Bruce of the University of the Philippines, Diliman; Drs. Bonita and Revilla of the College of Forestry, University of the Philippines, and staff of the FDC, Los Baños). In providing an overview of forest cover, all three surveys fit in quite well with the data discussed in chapter 3. At the regional and provincial levels, the results are difficult to reconcile.

It seems fair to conclude that the three visual interpretations of LANDSAT photomosaics in the Philippines have been at least as accurate with regard to national forest cover as the surveys done in the United States and Sweden. In fact, there are sound grounds for suggesting that the 1973 survey was inaccurate, and, more recently, the P-GFI seems to be superior to the 1987 SPOT survey (at least for the purposes of measuring actual deforestation).

The advantages of manual interpretation by Filipino photointerpreters are several: it is less costly, it can be done in-country, and it is conducted by people who are experts on Philippine forest cover. While the results may be accurate only for national and possibly regional forest totals, computerized analysis of remotely sensed data, at least in the Philippines, has yet to demonstrate that it is any more accurate.

At the same time, it must be remembered that neither manual nor computer interpretations of SPOT or LANDSAT imagery are a substitute for a traditional forest inventory. A forest inventory like the P-GFI can gather information on tree species, timber volume, and other forest products such as bamboo and rattan. Even though they are costly and time-consuming, forest inventories will still be necessary in the future.

In short, if the desired goal is a rough measure of forest cover at the national level, it is not clear that a computer-assisted analysis of remotely sensed data is better than manual interpretation of a photomosaic. If detailed information about tree species and volumes is required, then there is no substitute for a forest inventory involving ground samples.

This should not be interpreted as a criticism of remote sensing as a tool for measuring forest or vegetation cover; rather, it is an attempt to point out some of the difficulties involved in its application to the Philippines and, further, to raise the question of what is the ultimate goal of the data gathered in this manner. If the purpose is to integrate knowledge of forest cover into the planning process at the national, regional, or provincial level, then it must be judged to have been a complete failure. Knowledge of the extent of forest cover has yet to have any appreciable effect on controlling deforestation in the Philippines. The same could most likely be said for almost all tropical countries. In the Philippines, political consideration up till 1986 consistently overrode the utility of forestry data gathered by remote sensing.

One of the major difficulties of the 1987 SPOT survey was its use of forest/vegetation categories that were not consistent with previous and continuing work in the Philippines. Therefore, the results are not directly comparable with earlier forest inventories. This raises very serious questions regarding the utility of remote-sensing surveys in studying previous deforestation.

First, the use of different categories in different surveys means that it may be almost impossible to determine accurate rates of deforestation. Based on the Philippine experience with the 1987 SPOT survey, I suggest that future remote-sensing work must make a serious effort to ensure that results are as compatible as possible with previous work. If not, the work may be of limited value to understanding tropical deforestation. In addition, changing technologies and different sensors may make future comparisons difficult in and of themselves. Second, the existence of forest inventory results from different surveys, even if they are roughly comparable at the national level, can have profound implications for calculating rates of deforestation. For instance, if forest area was 10 million ha in 1980 and either 7 or 8 million ha in 1990, this means that deforestation was either 3.5% or 2.2% per annum. In this case, a 12% difference in forest area in 1990 leads to a 37% difference in the calculated rate of deforestation.

In order to provide data useful for the study of deforestation rates at a national scale, future remote-sensing forest surveys must (1) more fully integrate their results with government planning efforts and (2) ensure comparability with previous work. While I sympathize with the calls of Grainger (1983), Myers (1988c), and Woodwell et al. (1983) for increased remote sensing of tropical forests, it is with the important qualification that remote sensing will fulfill its potential role only if the above issues are addressed.

Several comments on the relationship between remote sensing and the contribution of tropical deforestation to global climate change seem appropriate. If tropical deforestation is contributing to the total CO_2 released into the atmosphere, and if global monitoring of tropical forests by remote sensing becomes a reality, then the results of this study may be of relevance.

In the case of the Philippines, it is obvious that the primary dipterocarp forests are almost completely gone (from approximately 10 million ha in the early 1950s to less than 1 million ha today). Since these forests have a great deal more biomass than mangrove, mossy, or secondary dipterocarp forests, it may be that the Philippines' contribution to atmospheric CO_2 was greater in the 1950–80 period than at present. In other words, the forests with the largest amounts of stored carbon have already been destroyed. Since deforestation has been extensive in Thailand, it may be true there also. Thus, the contribution of certain countries to global atmospheric CO_2 may actually have been declining in the recent past.

In addition, in the Philippines, knowledge of the changes of land use from primary to secondary forest, to agriculture, to open land, is weak. As a result, definitive statements regarding carbon fixation by vegetation growing on previously forested lands cannot be made (see Palm et al. 1986 for a discussion of the difficulties in estimating atmospheric CO_2 from deforestation in Southeast Asia, and Detwiler and Hall 1988 for the relationship between tropical forests and the global carbon cycle). Not only do we know very little about deforestation itself, but, in addition, we know very little about what happens after deforestation. Our lack of detailed understanding of the types of land use which occur after removal of forest cover hampers efforts to control deforestation and to model the global carbon cycle. In terms of data gathering, it is obvious that new survey instruments are required. An agricultural census every ten years should include information on abandoned agricultural land for the purposes of both deforestation analysis and climate modeling.

Lastly, if the purpose of monitoring tropical forest cover and improving climate models is to help reduce the contribution of tropical deforestation to increases in global atmospheric CO_2, and if the deforestation model in this study is correct for the Philippines, then it is difficult to be optimistic about solving the problem. The government leaders in the Philippines, Thailand, and Indonesia have shown very little concern for the environment of their own countries, let alone the world. If it can be proven that tropical deforestation is contributing to increased CO_2 levels and global environmental change, it is difficult to imagine how this would change the behavior of Third World leaders who benefit from the deforestation process.

Summary and Conclusion

Despite the difficulties posed by the quality of some of the available data sets, a detailed examination of deforestation in the postwar Philippines is possible. My study would have benefited greatly from similar studies for other nations; clearly the lack of such studies seriously hampers our understanding of tropical deforestation. In fact, given the attention that tropical deforestation has received recently, the lack of such studies is surprising.

Not only is our understanding of tropical deforestation incomplete but, just as important, our understanding of what happens after deforestation is incomplete. This is particularly the case with agriculture in the Philippines. Is deforestation followed by permanent or shifting cultivation? If shifting cultivation is the normal pattern, how long do farmers work a plot before it is abandoned or, alternatively, converted to open land? Are subsistence or commercial crops being grown? Our lack of knowledge regarding these topics has profound implications for understanding and solving the problem of deforestation and, also, for modeling global climate change. The lack of detailed studies about what happens after deforestation parallels the lack of studies about deforestation itself and both should be priority areas for future research.

An evaluation of previous quantitative work on deforestation and forest cover has revealed several major shortcomings. First, most studies have used percentage forest cover as the dependent variable and, therefore, cannot be considered to be studies of deforestation itself. Second, with two exceptions, all previous quantitative work has been cross-national and, as such, of relatively little value for designing policy instruments at the level of an individual country. Third, with the exception of the present study, no previous quantitative work has analyzed one country and had deforestation as the dependent variable.

The choice of the dependent variable is crucial. The higher explanatory power achieved by using forest cover as opposed to deforestation as the dependent variable indicates that the two dependent variables are not synonymous as most authors have assumed. The forest cover which remains today is the end result of a history of hundreds of years, but the deforestation of the past five years is the result of recent forces. As this study has indicated, a negative relationship between forest cover and population density does not mean that population density is the cause of deforestation. In short, most previous quantitative work on tropical deforestation, by having forest cover as the dependent variable, has not stated the problem correctly.

The analysis also indicates that there may be more uniqueness to country case histories of deforestation than is recognized in the literature. The Philippines differs from Thailand and Brazil in its archipelagic nature. The existence of large-scale commercial forests in Southeast Asia and the resulting commercial logging are important to explaining deforestation in that part of the world. While numerous other differences among countries could be set forth, it is clear that at least the type of forest cover and the physical nature of the country are important considerations in explaining deforestation. These individual features cannot be captured by cross-national studies.

The variables most important in the panel analysis of deforestation between 1970 and 1980 were ACC in 1970 and the increase in agricultural area. In particular, the increase in population was not important in explaining the variation in deforestation across provinces. This study is the first quantitative examination of tropical deforestation to conclude that population pressure is not an important explanatory variable of deforestation. An important topic for future research is whether or not population pressure is a significant explanatory variable of deforestation in countries other than the Philippines. This will of necessity require detailed case studies on a nation-by-nation basis.

I have emphasized the importance of elite control and corruption in explaining deforestation in the Philippines. It has worked in numerous ways: first, it facilitated the granting of primary forest resources to a small group of logging concessionaires; second, it led to corrupt and inefficient government regulation of the logging process; third, it encouraged migration of the poor to previously forested areas in an attempt to preclude structural reform of the socioeconomic system; fourth, it led to deliberate manipulation of government data on forest cover designed to mislead the Filipino media and forestry community and foreign researchers; fifth, the destructive logging set the tone for the poor migrants who followed and fueled the process of forest destruction; sixth, by concentrating financial returns in the hands of the elite, logging exacerbated the unequal distribution of income, which is most likely the greatest structural problem faced by the Philippines. As a result of the above factors, the destruction of the Philippine forest has not led to development in any meaningful sense of the word.

The most important feature of the scenario painted above is that the Philippine government had a large amount of control over the process and deliberately turned this control over to a small group of people. The process did not just happen; rather, it served the financial interests of the wealthy and well-connected. Unfortunately, the history of the Philippine forests in the postwar period is an apt metaphor for Philippine development during this period, as the widespread poverty of the Filipino people today demonstrates. A corollary to the fact that population pressure was not a major cause of deforestation in the 1970–80 period is that the Philippine government had more control over the forest resource than is commonly assumed. In other words, large-scale deforestation was not inevitable.

The history of deforestation in the postwar Philippines is to be contrasted with that of Malaysia. While deforestation has been extensive in Malaysia, at least productive, commercial agriculture involving rubber, palm oil, cocoa, and pepper has followed and, partly as a result, Malaysia now enjoys a per capita income at least three times that of the Philippines. Indonesia, like the Philippines, is an archipelago and has extensive commercial forests controlled by a small group of people. Possibly the best that can be said of the Philippine experience with deforestation is that it provides a good example of how a country can impoverish itself while benefiting a small group of people. This is precisely the direction that Indonesia appears to be taking with its forests.

In conclusion, I would like to note that the process of deforestation described for the Philippines is not amenable to a technical solution. The major questions do not concern the relative merits of different silvicultural techniques or the appropriate rate of discount to be used, although these are important issues. Rather, the fundamental issue is, who has the right to use the forest resource? In the Philippines, the answer has invariably been that the forest belongs to loggers and their allies. In short, deforestation is most appropriately studied by a multidisciplinary approach which emphasizes the socioeconomic and political environment in which the actual process of deforestation occurs.

Bibliography

Abad, Ricardo G. 1981. Internal Migration in the Philippines: A Review of Research Findings. *Philippine Studies* 29:129–43.

Abejo, Socorro. 1985. Migration to and from the National Capital Region: 1975–1980. *Journal of Philippine Statistics* 36:ix-xxii.

Agaloos, Bernardo C. 1984. Silvicultural and Logging Systems in the Philippines. In *Proceedings of the First ASEAN Forestry Congress* 2:210–33. Quezon City: Bureau of Forest Development.

_____. 1976. Aerial Photography in Forest Surveys. In *Asian Forestry Industry Yearbook*, pp. 33–36.

_____. 1965a. *Forest Resource Statistics for Eastern Mindanao.* Manila: Bureau of Forestry.

_____. 1965b. *Forest Resource Statistics for Mindanao.* Manila: Bureau of Forestry.

_____. 1964a. *Forest Resource Statistics for Central Mindanao.* Manila: Bureau of Forestry.

_____. 1964b. *Forest Resource Statistics for Western Mindanao.* Manila: Bureau of Forestry.

Agaloos, Bernardo C., and Santos, Teofilo A. 1968. *Forest Resources of Palawan.* Manila: Bureau of Forestry.

Agricultural Policy and Strategy Team. 1986. *Agenda for Action for the Philippine Rural Sector.* Los Baños: University of the Philippines, Los Baños.

Agtani, T. C. 1964. Kainginism: Philippine Forestry's No. 1 Enemy. *Philippine Lumberman* 10:21.

Aiken, Robert S., and Moss, Michael R. 1975. Man's Impact on the Tropical Rainforest of Peninsular Malaysia: A Review. *Biological Conservation* 8:213–29.

Aki, Koichi, and Berthelot, R. 1974. Hydrology of Humid Tropical Asia. In UNESCO, *Natural Resources of Humid Tropical Asia.* Paris: UNESCO.

Alano, Bienvenido P. 1984. Import Smuggling in the Philippines: An Economic Analysis. *Journal of Philippine Development* 11:157–90.

Alcala, Angel C. 1987. The Conservation of Our Marine Resources. *Solidarity* 115:89–91.

Allen, Julia C., and Barnes, Douglas F. 1985. The Causes of Deforestation in Developing Countries. *Annals of the Association of American Geographers* 75:163–84.

Allied Geographical Section: Southwest Pacific Area. 1944. *Timber Resources of the Philippine Islands.* n.p.

Amos, Felipe R. 1954. *Forest Resources of the Philippines.* Manila: Bureau of Forestry.

Andal, Sergio M. 1986. Can Dollar Salting and Technical Smuggling in the Garments Industry Be Controlled? Center for Research and Communication Staff Memos no. 16: 2–6.

Anderson, Dennis. 1986. Declining Tree Stocks in African Countries. *World Development* 14:853–64.

Anderson, James N. 1987. Lands at Risk, People at Risk: Perspectives on Tropical Forest Transformations in the Philippines. In *Lands at Risk in the Third World: Local-Level Perspectives*, edited by Peter D. Little, Michael M. Horowitz, and Endre Nyerges. Boulder: Westview Press.

_____. 1982. Rapid Rural "Development": Performances and Consequences in the Philippines. In *Too Rapid Rural Development: Perceptions and Perspectives from Southeast Asia*, edited by Colin MacAndrews and Chia Lin Sien. Athens: Ohio University Press.

Angeles, Marian. See Segura-de los Angeles, Marian.

Anti-Slavery Society. 1983. *The Philippines: Authoritarian Government, Multinationals and Ancestral Lands*. London: Anti-Slavery Society.

Aquino, Belinda A. 1987. *Politics of Plunder*. Quezon City: University of the Philippines, College of Public Administration.

Aranas, Armando L. 1973. Photogrammetry: A Modern Tool of Forestry. In *First National Agricultural System Research Congress: Forest Production Research Papers*. Los Baños: Philippine Council for Agricultural Research.

Asia Society. 1986. *The Philippines: Facing the Future*. New York: Asia Society.

Asian Development Bank. 1987a. *Philippines: Environmental and Natural Resources Briefing Profile*. Manila: Asian Development Bank.

_____. 1987b. *A Review of Forestry and Forest Industries in the Asia-Pacific Region*. Manila: Asian Development Bank.

_____. 1976. *The Forest Economy of the Philippines*. Manila: Asian Development Bank.

Asiaweek. 1989. The High Price of Conservation. 15 (10): 31 (March 10).

_____. 1988. Rape of the Rain Forest. 14 (48): 52–60 (Nov. 25).

Astillero, Emanuel I. 1976. The Bicol River Basin Development Program: An Exercise in Integrated River Basin Planning. *Philippine Planning Journal* 7:37–43.

Ayub, Arshad. 1979. National Agricultural Policy and Its Implications on Forest Development in the Country. *Malaysian Forester* 42:348–53.

Bahrein, Tunku Shamsul. 1968. Land Conflicts in the Tanay Resettlement Project (Rizal), Philippines. *Journal of Tropical Geography* 27:50–58.

Bajracharya, Deepak. 1983. Fuel, Food, or Forest? Dilemmas in a Nepali Village. *World Development* 11:1057–74.

Barney, G. O. 1980. *Entering the Twenty First Century: The Global 2000, Report to the President*. Washington, D.C.: Council on Environmental Quality.

Basa, Virgilio F. 1982. Land Classification in the Philippines. In *Proceedings: Symposium on Land Use in Upland Areas*. Bulletin no. 89. Manila: National Research Council of the Philippines.

Bedard, Paul W. 1958. Reconnaissance, Classification, and Mapping of Philippine Forests. In *Study of Tropical Vegetation: Proceedings of the Kandy Symposium, March 19–21*. Paris: UNESCO.

Bee, Ooi Jin. 1987. *Depletion of the Forest Resources in the Philippines*. Field Report Series no. 18. Singapore: Institute of Southeast Asian Studies.

Beets, Nico; Neggers, Jan; DeLa Vega, Angelito; and Zevenbergen, William. 1986. *Low External Input Agriculture*. Manila: Netherlands Organization for International Development Cooperation/Filippijnen.

Bello, Walden. 1988. From Dictatorship to Elite Populism: The United States and the Philippine Crisis. In *Crisis and Confrontation: Ronald Reagan's Foreign Policy*, edited by Morris H. Morley. Totowa: Rowman and Littlefield.

Belsky, Jill M. 1989. Household Food Security, Land Use, and Agroforestry: A Comparative Study in the Philippines and Indonesia. Paper presented at the Annual Meeting of the Rural Sociological Society, Seattle.

Belsky, Jill M., and Siebert, Steve. 1985. Social Stratification, Agricultural Intensification, and Environmental Degradation in Leyte, Philippines: Implications for Sustainable Development. Paper presented at Sustainable Development of Natural Resources in the Third World—An International Symposium, Ohio University, Athens.

Bennagen, M. Eugenia C. 1982. *Staple Food Consumption in the Philippines*. Working Paper no. 5. Washington, D.C.: International Food Policy Research Institute.

Bernardo, Thelma S. 1982–83. Historical Analysis of Inter-Regional Migration in the Philippines. *Central Luzon State University Scientific Journal* 3:68–97.

Bhumibhamon, Suree. 1986. *The Environmental and Socio-Economic Aspects of Tropical Deforestation: A Case Study of Thailand*. Bangkok: Kasetsart University, Faculty of Forestry, Department of Silviculture.

Binswanger, Hans P. 1989. *Brazilian Policies That Encourage Deforestation in the Amazon*. Environment Department Working Paper no. 16. Washington, D.C.: World Bank.

Blaikie, Piers. 1985. *The Political Economy of Soil Erosion in Developing Countries*. London: Longman.

Blaikie, Piers, and Brookfield, Harold, eds. 1987. *Land Degradation and Society*. London: Methuen.

Blanche, Catalino A. 1975. An Overview of the Effects and Implications of Philippine Selective Logging on the Forest Ecosystem. In *Proceedings of the Symposium on the Long-Term Effects of Logging in Southeast Asia, Bogor, June 24–27*, edited by R. S. Suparto. Special Publication no. 3. Bogor, India: BIOTROP.

Boado, Eufresina L. 1988. Incentive Policies and Forest Use in the Philippines. In *Public Policies and the Misuse of Forest Resources*, edited by Robert Repetto and Malcolm Gillis. Cambridge: Cambridge University Press.

Bonita, Manuel L. 1977. Location of Forest Industries in the Philippines. In *Report of the FAO/ Norway Seminar on Storage, Transport, and Shipping of Wood*. Rome: FAO.

Bonita, Manuel L., and Revilla, Adolfo. 1977. The Philippine Forest Resources, 1976–2026. In *PREPF*, vol. 2, *Project Reports and Technical Papers*. Manila: Development Academy of the Philippines.

Bonner, Raymond. 1987. *Waltzing with a Dictator: The Marcoses and the Making of American Policy*. New York: Times Books.

Borja, Luis J. 1929. The Philippine Lumber Industry. *Economic Geography* 5:194–202.

Bowonder, B. 1987. Environmental Problems in Developing Countries. *Progress in Physical Geography* 11:246–59.

_____. 1985–86. Deforestation in Developing Countries. *Journal of Environmental Systems* 15: 171–92.

_____. 1982. Deforestation in India. *International Journal of Environmental Studies* 18:223–36.

Briones, Nicomedes D. 1986. Estimating Erosion Costs: A Philippine Case Study in the Lower Agno River Watershed. In *Watershed Resources Management*, edited by William K. Easter, John A. Dixon, and Maynard M. Hufschmidt. Boulder: Westview Press.

Bruce, Romeo C. 1987. Status of Land Degradation in the Philippines. Paper presented at the University of the Philippines, Diliman.

_____. 1977. *Save Our Forests Today and Live Better Tomorrow*. Manila: Development Academy of the Philippines.

Bulatao, Rodolfo A., ed. 1976. *Philippine Population Research*. Manila: Population Center Foundation.

Bureau of Agricultural Economics. 1985. *Bicol Region Profile*. Quezon City: Mininstry of Agriculture and Food.

Bureau of Census and Statistics. 1963. *Census of the Philippines, 1960: Agriculture*. Manila: Department of Commerce and Industry.

_____. 1957. *Yearbook of Philippine Statistics*. Manila: Bureau of Census and Statistics.

_____. 1952. *Summary Report on the 1948 Census of Agriculture*. Manila: Bureau of Printing.

Bureau of Forestry. 1902. *Report of the Bureau of Forestry of the Philippine Islands, 7/1/1901–9/1/1902*. Manila: Philippine Commission.

Bureau of Forest Development. See Forest Management Bureau.

Bureau of Soils. 1977. *Land Capability Classes*. Manila: Department of Environment and Natural Resources.

Burgess, P. F. 1973. The Impact of Commerical Forestry on the Hill Forests of the Malay Peninsula. In *Proceedings of the Symposium on Biological Resources and National Development*, edited by E. Soepadmo and K. G. Singh. Kuala Lumpur: Malayan Nature Society.

_____. 1971. The Effect of Logging on Hill Dipterocarp Forests. *Malayan Nature Journal* 24: 231–37.

Byron, Neil, and Waugh, Geoffrey. 1988. Forestry and Fisheries in the Asian-Pacific Region: Issues in Natural Resource Management. *Asian-Pacific Economic Literature* 2:46–80.

Callaham, R. Z., and Buckman, R. E. 1981. *Some Perspectives of Forestry in the Philippines, Indonesia, Malaysia, and Thailand*. Washington, D.C.: U.S. Department of Agriculture/Forest Service.

Cant, Garth. 1979. The Philippines: Spatial Patterns and Spatial Planning. In *Migration and Development in South-East Asia: A Demographic Perspective*, edited by Robin J. Pryor. Kuala Lumpur: Oxford University Press.

Capistrano, Ana Doris, and Marten, Gerald G. 1986. Agriculture in Southeast Asia. In *Traditional Agriculture in Southeast Asia: A Human Ecology Perspective*, edited by Gerald G. Marten. Boulder: Westview Press.

Capistrano, Ana Doris, and Fujisaka, Sam. 1984. *Tenure, Technology, and Productivity of Agroforestry Schemes*. Working Paper 84–06. Manila: Philippine Institute of Development Studies.

Carbonell-Catilo, Aurora. 1986. The Philippines: The Politics of Plunder. *Corruption and Reform* 1:235–43.

Carino, Ledivina V., ed. 1986. *Bureaucratic Corruption in Asia*. Manila: University of the Philippines, College of Public Administration.

Carroll, John J. 1983. Agrarian Reform, Productivity, and Equity: Two Studies. In *Second View from the Paddy*, edited by Antonio J. Ledesma, Perla Q. Makil, and Virginia A. Miralao. Manila: Ateneo de Manila University Press.

Caufield, Catherine. 1985. *In the Rainforest*. New York: Alfred Knopf.

Census Office of the Philippine Islands. 1920. *Census of the Philippine Islands, 1918*. Manila: Bureau of Printing.

Certeza and Co. 1988. Personal communication.

Chuan, Goh Kim. 1982. Environmental Impact of Economic Development in Peninsular Malaysia: A Review. *Applied Geography* 2:3–16.

Clark, Colin W. 1973. The Economics of Overexploitation. *Science* 181:630–34.

Clark, David. 1988. Review Article: Recent Literature on Corruption. *Asian Journal of Public Administration* 10:120–25.

Concepcion, Mercedes B., ed. 1983. *Population of the Philippines: Current Perspectives and Future Prospects.* Manila: National Economic Development Authority.

Conklin, Harold C. 1957. *Hanunoo Agriculture in the Philippines.* Forestry Development Paper no. 12. Rome: FAO.

Coppin, P. 1984. *A Guideline for Photo Interpretation and Thematic Forest Mapping,* part 1, *The Luzon Area.* Forest Resources Inventory Report no. 4. Quezon City: Bureau of Forest Development.

Coppin, P., and Lennertz, R. 1986. *A Methodology for the Establishment of a Geographically Based, Multi-Temporal, and Multi-Thematic Land Cover Information System for Forest Resources Management in the Philippines.* Forest Resources Inventory Report no. 8. Quezon City: Bureau of Forest Development.

Cornista, Luzviminda B; Javier, Filomena A.; and Escueta, Eva F. 1986. *Land Tenure and Resource Use among Upland Farmers.* Paper Series no. 2. Los Baños: Agrarian Reform Institute.

Costello, Michael A. 1984. Social Change in Mindanao: A Review of the Research of a Decade. *Kinadman* 6:1–41.

Cressy, Paul F. 1960. Urbanization in the Philippines. *Sociology and Social Research* 44:402–9.

Crone, Donald. 1986. ASEAN's Third Decade: Building Greater Equity. Paper presented at the Annual Meeting of the Association of Asian Studies, Chicago, March.

Cruz, Cerenilla A., and Segura-de los Angeles, Marian. 1984. *Policy Issues on Commercial Forest Management.* Working Paper 84–03. Manila: Philippine Institute for Development Studies.

Cruz, M. Concepcion. 1990. Personal communication.

_____. 1988. Upland Population and Migration Patterns: Estimation Methods and Policy Issues. Paper presented at the Upland Resource Policy Conference sponsored by the Philippine Institute for Development Studies and the Department of Environment and Natural Resources, Quezon City, March 14.

_____. 1984. Population Pressure, Migration, and Markets: Implications for Upland Development. Paper presented at the Workshop on Economic Policies for Forest Resources Management sponsored by the Philippine Institute for Development Studies, Los Baños, February 17–18.

Cruz, M. Concepcion, and Zosa-Feranil, Imelda. 1988. Policy Implications of Population Pressure in Philippine Uplands. Paper prepared for the World Bank/Canadian International Development Agency Study on Forestry, Fisheries, and Agriculture Resource Management.

Cruz, M. Concepcion; Zosa-Feranil, Imelda; and Goce, Cristela L. 1986. *Population Pressure and Migration: Implications for Upland Development in the Philippines.* Working Paper 86–06. Los Baños: Center for Policy and Development Studies.

Dacanay, Placido. 1943. *The Forest Resources of the Philippines.* Manila: Bureau of Forestry and Fishery.

Daniel, J. G., and Kulasingam, A. 1974. Problems Arising from Large Scale Forest Clearing for Agricultural Use: The Malaysian Experience. *Malaysian Forester* 37:152–60.

Dasmann, Raymond F.; Milton, John P.; and Freeman, Peter H. 1979. *Ecological Principles of Economic Development.* New York: John Wiley.

Data Assessment and Review Team. 1983. *Regional Statistical and Agricultural Profile,* vol. 4, *Southern Tagalog.* Quezon City: Ministry of Agriculture.

1987. *Philippines: Food Policy in Transition.* Agricultural Economic Department Paper no. 87–03. Los Baños: International Rice Research Institute.

———. 1983. *Economic Policies and Philippine Agriculture.* Manila: Philippine Institute for Development Studies, Working Paper 83–02.

———. 1982. The Impact of Economic Policies on Agricultural Incentives. Paper presented at the Development Academy of the Philippines, October 6.

David, Cristina C., and Barker, Randolph. 1979. Agricultural Growth in the Philippines, 1948–1971. In *Agricultural Growth in Japan, Taiwan, Korea, and the Philippines,* edited by Yujiro Hayami, Vernon W. Ruttan, and Herman M. Southworth. Honolulu: University of Hawaii Press.

DeBeer, Jenne H., and McDermott, Melanie J. 1989. *The Economic Value of Non-Timber Forest Products in Southeast Asia.* Amsterdam: Netherlands Committee for IUCN.

DeDios, Emmanuel S., ed. 1984. *An Analysis of the Philippine Economic Crisis.* Quezon City: University of the Philippines Press.

Department of Agrarian Reform. 1988. Personal communication.

———. 1987. *Settlement Projects Administered by the Department of Agrarian Reform as of 12/31/87.* Quezon City: Bureau of Resettlement.

Department of Agriculture. 1988. *Provincial Profiles: Philippines.* Quezon City: Department of Agriculture.

Department of Agriculture and Natural Resources. n.d. *Crop, Livestock, and Natural Resources Statistics, 1958 and 1959.* Manila: Agricultural Economics Division.

———. n.d. *Crop and Livestock Statistics, 1956 and 1957.* Manila: Agricultural Economics Division.

———. n.d. *Crop and Livestock Statistics, 1954 and 1955.* Manila: Agricultural Economics Division.

Department of Environment and Natural Resources, Ford Foundation, University of the Philippines, Los Baños, Foundation and Forestry Development Center. 1987. *Policy and Program Agenda for the Environment and Natural Resources Sector: Integrated Report.* Los Baños: Forestry Development Center.

Department of Public Works and Highways. 1988. Existing Road Length by System Classification and Standard, 1965–87. Computer printouts and hand-copied reports.

Detwiler, R. P., and Hall, Charles. 1988. Tropical Forests and the Global Carbon Cycle. *Science* 239:42–47.

Diaz, Celso P. 1982. Socio-Economic Thrusts in an Integrated Forest-Management System: The Philippine Case. In *Socio-Economic Effects and Constraints in Tropical Forest Management,* edited by E. G. Hallsworth. London: John Wiley and Sons.

Dillon, William R., and Goldstein, Matthew. 1984. *Multivariate Analysis: Methods and Applications.* New York: John Wiley.

Doeppers, Dan. 1986. Personal communication.

Dove, Michael R. 1983. Theories of Swidden Agriculture and the Political Economy of Ignorance. *Agroforestry Systems* 1:85–99.

Dowling, J. Malcolm, and Soo, David. 1983. *Income Distribution and Economic Growth in Developing Countries.* Economic Staff Paper no. 15. Manila: Asian Development Bank.

Duckham, A. N., and Masefield, G. B. 1969. *Farming Systems of the World.* New York: Praeger.

Dugan, Pat. 1987. In Seminar on Managing and Conserving Our Environmental Resources, moderated by F. Sionil Jose. *Solidarity* 115:3–42.

Durst, Patrick B. 1985. *Factors Influencing Discrepancies between Forest Products Trade Reports of the Philippines and Its Trading Partners.* Policy Paper no. 17. Los Baños: Forestry Development Center.

Eckholm, Eric. 1979. *Planting for the Future: Forestry for Human Needs.* Paper 26. Washington, D.C.: Worldwatch.

_____. 1976. *Losing Ground: Environmental Stress and World Food Prospects.* New York: Norton.

Economist. 1989. A Brazilian Tale. 310 (7590): 31–32 (Feb. 18).

Economist Intelligence Unit. 1989. *Philippines: Country Report.* London: Economist Intelligence Unit.

Eder, James F. 1989. Review of *Depletion of Forest Resources in the Philippines,* by Ooi Jin Bee. *Journal of Asian Studies* 48:679–81.

_____. 1977. Agricultural Intensification and the Returns to Labour in the Philippine Swidden System. *Pacific Viewpoint* 18:1–21.

Edgerton, Ronald K. 1983. Social Disintegration on a Contemporary Philippine Frontier: The Case of Bukidnon, Mindanao. *Journal of Contemporary Asia* 13:151–75.

Egerton, J. O. 1953. Notes on Logging in the Philippines. *Malayan Forester* 16:146–56.

Episcopal Commission of Tribal Filipinos. 1982. *Tribal Forum* 3 (6): entire issue.

Fearnside, Philip M. 1986. Spatial Concentration of Deforestation in the Brazilian Amazon. *Ambio* 15:74–81.

_____. 1985. Environmental Change and Deforestation in the Brazilian Amazon. In *Change in the Amazon Basin: Man's Impact on Forests and Rivers,* edited by J. Hemming. Manchester: Manchester University Press.

_____. 1982. Deforestation in the Brazilian Amazon: How Fast Is It Occurring? *Interciencia* 7:82–88.

Feder, Ernest. 1983. *Perverse Development.* Quezon City: Foundation for Nationalist Studies.

Feder, Gershon; Onchan, T.; and Chalamwong, Y. 1988. Land Policies and Farm Performance in Thailand's Forest Reserve Areas. *Economic Development and Cultural Change* 36:483–501.

Feeny, David. 1984. *Agricultural Expansion and Forest Depletion in Thailand, 1900–1975.* Economic Growth Center Discussion Paper no. 458. New Haven: Yale University.

Finney, C. E., and Western, S. 1986. An Economic Analysis of Environmental Protection and Management: An Example from the Philippines. *Environmentalist* 6:45–61.

Fleiger, Wilhelm; Koppin, Brigada; and Lim, Carmencita. 1976. *Geographical Patterns of Internal Migration in the Philippines.* Manila: National Economic Development Authority and National Census and Statistics Office.

Flores, Jamil Maidan. 1986. A Call for Heroism. *Philippine Panorama* 15 (31): 5, 21 (Aug. 24).

Food and Agriculture Organization. 1987. *Assessment of Forest Resources in Six Countries.* Special Study on Forest Management, Afforestation, and Utilization of Forest Resources in the Developing Regions (Asia-Pacific Region)—Field Document 17. Bangkok: FAO.

_____. 1963. *World Forest Inventory.* Geneva: FAO.

_____. 1960. *World Forest Inventory.* Rome: FAO.

_____. 1955. *World Forest Resources.* Rome: FAO.

_____. 1948. *Forest Resources of the World.* Washington, D.C.: FAO.

_____. 1946. *Forestry and Forest Products: World Situation, 1937–1946.* Washington, D.C.: FAO.

Food and Agricultural Organization/United Nations Environment Programme. 1982. *Tropical Forest Resources Assessment Project.* 4 vols. Rome: FAO.

_____. 1981. *Forest Resources of Tropical Asia.* Rome: FAO.

Forest Management Bureau. 1988. *Natural Forest Resources of the Philippines.* Manila: Philippine-German Forest Resources Inventory Project.

_____. 1986–88. *Forest Resources of Region 1–Region 12.* 12 vols. Manila: P-GFRI Project.

_____. 1968–87. *Philippine Forestry Statistics* (issued annually). Manila: FMB.

Forestry Development Center. 1985. *A 50-Year Development Program for the Philippines.* Los Baños: FDC.

French, David. 1986. Confronting an Unsolvable Problem: Deforestation in Malawi. *World Development* 14:531–40.

Fuchs, Roland J., and Luna, Telesforo W. 1972. *Spatial Patterns of Socio-Economic Structure and Change in the Philippines, 1939–1960.* Working Paper no. 26. Honolulu: East-West Population Institute.

Fujisaka, Sam. 1986. Pioneer Shifting Cultivation, Farmer Knowledge, and an Upland Ecosystem: Co-Evolution and Systems Sustainability in Calminoe, Philippines. *Philippine Quarterly of Culture and Society* 14:137–64.

Gamser, Matthew S. 1980. The Forest Resource and Rural Energy Development. *World Development* 8:769–80.

Ganapin, Delfin J. 1987. Forest Resources and Timber Trade. *Solidarity* 115:53–64.

Gastellu-Etchegorry, J. P., and Sinulingga, A. B. 1988. Designing a GIS for the Study of Forest Evolution in Central Java. *Tijdschrift voor Economische en Sociale Geografie* 79:93–103.

General Electric Co. and Department of Natural Resources. 1977. *Forest Inventory of the Philippine Islands Using LANDSAT Multispectral Scanner Digital Data.* Beltsville, Md.: General Electric Co.

Generalao, Maximino L. 1975. The Complex Problem of Deforestation in the Philippines. *Philippine Lumberman* 21:8–11.

Gentry, A. H., and Lopez-Parodi, J. 1980. Deforestation and Increased Flooding of the Upper Amazon. *Science* 210:1354–56.

Gill, Tom. 1960a. Forestry Proposals for the Philippines. A Report to the International Cooperation Administration and the National Economic Council.

_____. 1960b. What Is Happening to Philippine Forests? *Philippine Journal of Forestry* 16:17–30.

Gillis, Malcolm. 1988. The Logging Industry in Tropical Asia. In *People of the Tropical Rain Forest,* edited by Julie Sloan Denslow and Christine Padoch. Berkeley: University of California Press.

Gomez-Pompa, A.; Vazquez-Yanes, C.; and Guevara, S. 1972. The Tropical Rain Forest: A Non-renewable Resource. *Science* 177:762–65.

Gooch, Winslow L. 1953. *Forest Industries of the Philippines.* Manila: Bureau of Forestry/U.S. Mutual Security Agency.

Gorra, Marilyn N. 1986. Social Development Alternatives in the Philippines. *Regional Development Dialogue* 7:136–60.

Gowing, Peter G. 1980. Contrasting Agenda for Peace in the Muslim South. *Philippine Quarterly of Culture and Society* 8:286–302.

Grainger, Alan. 1989. Personal communication.

_____. 1987. A Land Use Simulation Model for the Humid Tropics. Paper presented at the Land and Resource Evaluation for National Planning in the Tropics International Conference and Workshop, Chetumal, Mexico, Jan. 25–31.

_____. 1986. The Future Role of the Tropical Rain Forests in the World Forest Economy. Ph.D. diss., Oxford University.

_____. 1983. Improving the Monitoring of Deforestation in the Humid Tropics. In *Tropical Rain Forest: Ecology and Management*, edited by S. L. Sutton, T. C. Whitmore, and A. C. Chadwick. Oxford: Blackwell.

_____. 1980. The State of the World's Tropical Forests. *Ecologist* 10:6–54.

Grimwood, I. R. 1975. *National Parks and Wildlife Conservation in the Philippines.* Rome: FAO.

Gulcur, Macid Y. 1968. Renewable Natural Resources and Their Problems in the Philippines. In *Conservation in Tropical South East Asia*, edited by Lee M. Talbot and Martha H. Talbot. Morges: IUCN.

Guppy, Nicholas. 1984. Tropical Deforestation: A Global View. *Foreign Affairs* (Spring): 928–65.

Gwyer, G. 1978. Developing Hillside Farming Systems for the Humid Tropics: The Case of the Philippines. *Oxford Agrarian Studies* 7:1–37.

_____. 1977. *Agricultural Employment and Farm Incomes in Relation to Land Classes: A Regional Analysis.* Regional Planning Assistance Project, Technical Paper no. 6. Manila: NEDA-UNDP/IBRD.

Hackenberg, Robert, and Hackenberg, Beverly H. 1971. Secondary Development and Anticipatory Urbanization in Davao, Mindanao. *Pacific Viewpoint* 12:1–19.

Hainsworth, Geoffrey B. 1979. Economic Growth and Poverty in Southeast Asia: Malaysia, Indonesia, and the Philippines. *Pacific Affairs* 52:5–41.

Hainsworth, Reginald G., and Moyer, Raymond T. 1945. *Agricultural Geography of the Philippines.* Washington, D.C.: U.S. Department of Agriculture, Office of Foreign Agricultural Relations.

Hamilton, A. C. 1984. *Deforestation in Uganda.* Nairobi: Oxford University Press.

Hamilton, Larry S. 1988. What's to Blame for Floods in Bangladesh? *Centerviews* 6:5.

_____. 1986. *Towards Clarifying the Appropriate Mandate in Forestry for Watershed Rehabilitation and Management.* Environment and Policy Institute Reprint no. 94. Honolulu: East-West Center.

_____. 1985. *Overcoming Myths about Soil and Water Impacts of Tropical Forest Land Uses.* Environment and Policy Institute Reprint no. 86. Honolulu: East-West Center.

_____. 1984. *A Perspective on Forestry in Asia and the Pacific.* Environment and Policy Institute Reprint no. 67. Honolulu: East-West Center.

_____. 1982. Response of Tropical Forest Watersheds to Various Uses or Conversions. In *Proceedings of a Workshop on an Ecological Basis for Rational Resource Utilization in the Humid Tropics of Southeast Asia, Universiti Pertanian, Malaysia, January 18–22.*

Hayami, Yujiro; David, Cristina; Flores, Piedao; and Kikuchi, Masao. 1976. Agricultural Growth against a Land Resource Constraint: The Philippine Experience. *Australian Journal of Agricultural Economics* 20:144–59.

Headland, Thomas N. 1987. Ecosystemic Change in a Philippine Tropical Rainforest and Its Effect on a Negrito Foraging Society. Paper presented at the Conference on World Environmental History, Duke University, April 30-May 2.

Herrin, Alejandro N. 1985. Migration and Agricultural Development in the Philippines. In *Urbanization and Migration in ASEAN Development*, edited by Philip M. Hauser, Daniel B. Suits, and Nashiro Ogawa. Tokyo: National Institute for Research Advancement.

Hewitt, Kenneth. 1983. The Idea of Calamity in a Technocratic Age. In *Interpretations of Calamity*, edited by Kenneth Hewitt. Boston: Allen and Unwin.

Hicks, George L., and McNicoll, Geoffrey. 1971. *Trade and Growth in the Philippines.* Ithaca: Cornell University Press.

Hill, Hal, and Jayasuriya, Sisira. 1984. Philippine Economic Performance in Regional Perspective. *Contemporary Southeast Asia* 6:135–58.

Hirsch, Philip. 1987. Deforestation and Development in Thailand. *Singapore Journal of Tropical Geography* 8:129–38.

Hodgson, Gregor. 1988. Sedimentation Damage to Coral Reefs Due to Coastal Logging. Ph.D. diss., University of Hawaii.

Hong, Evelyne. 1987. *Natives of Sarawak: Survival in Borneo's Vanishing Forests.* Kuching: Institut Masyarakat.

Hooley, Richard, and Ruttan, Vernon W. 1969. The Philippines. In *Agricultural Development in Asia*, edited by R. T. Shand. Berkeley: University of California Press.

Houghton, Richard A., and Woodwell, George M. 1989. Global Climate Change. *Scientific American* 260:36–44.

Hosier, Richard H., and Milukas, Matthew V. 1989. Woodfuel Markets in Africa: Depletion or Development? Paper presented at the Annual Meeting of the Association of American Geographers, Baltimore, March 19–22.

Huke, Robert. 1986. Personal communication.

———. 1982. *Southeast Asia: Agroclimatic Map.* Los Baños: International Rice Research Institute.

———. 1963. Mindanao . . . Pioneer Frontier? *Philippine Geographical Journal* 7:74–83.

Hutacharoen, Malee. 1987. Application of Geographic Information Systems Technology to the Analysis of Deforestation and Associated Environmental Hazards in Northern Thailand. In *Proceedings of the Second Annual International Conference, Exhibits, and Workshops on Geographic Information Systems, San Francisco, October 26–30.*

Hyde, William F., and Hamilton, Lawrence S. 1988. Loss of Forest and Fuel-Wood Prices. *Environmental Conservation* 15:66–67.

Hyman, Eric L. 1984. Providing Public Lands for Smallholder Agroforestry in the Province of Ilocos Norte, Philippines. *Journal of Developing Areas* 18:177–90.

———. 1983. Forestry Administration and Policies in the Philippines. *Environmental Management* 7:511–24.

Institute of Population Studies. 1981. *Migration in Relation to Rural Development: ASEAN Level Report.* Bangkok: Chulalongkorn University.

Institute of Social Analysis. 1989. *Logging against the Natives of Sarawak.* Selangor: Institute of Social Analysis.

International Bank for Rural Development. 1972. Tropical Hardwood Trade in the Asia-Pacific Region: Issues and Opportunities. Economic Staff Working Paper no. 136.

International Labour Office. 1974. *Sharing in Development: A Programme of Employment, Equity, and Growth for the Philippines.* Geneva: ILO.

Ives, J., and Pitt, D. C., eds. 1988. *Deforestation: Social Dynamics in Watersheds and Mountain Ecosystems.* London: Routledge.

James, William E. 1983. *Asian Agriculture in Transition: Key Policy Issues.* Economic Staff Paper no. 19. Manila: Asian Development Bank.

Janzen, Daniel H. 1973. Tropical Agroecosystems. *Science* 182:1212–19.

Johnson, Nels, and Alcorn, Janis. 1989. *Ecological, Economic, and Development Values of Biological Diversity in Asia and the Near East.* Washington, D.C.: U.S. Agency for International Development, Bureau for Asia and the Near East.

Jones, Gavin W. 1983. *Structural Change and Prospects for Urbanization in Asian Countries.* Population Institute Paper no. 88. Honolulu: East-West Center.

Keith, H. G. 1956. *Report to the Government of the Philippines on Forest Policy and Legislation.* Report no. 470. Rome: FAO.

Kennedy, Peter. 1979. *A Guide to Econometrics.* Cambridge, Mass.: MIT Press.

Kerkvliet, Benedict J. 1974. Land Reform in the Philippines since the Marcos Coup. *Pacific Affairs* 47:286–304.

Khan, Azizur Rahman. 1977. *Growth and Inequality in the Rural Philippines.* In International Labour Office, *Poverty and Landlessness in Rural Asia.* Geneva: ILO.

Kikuchi, Masao, and Hayami, Yujiro. 1978. Agricultural Growth against a Land Resource Constraint: A Comparative History of Japan, Taiwan, Korea, and the Philippines. *Journal of Economic History* 38:839–64.

Kim, Jun. 1972. Net Internal Migration in the Philippines, 1960–1970. *Journal of Philippine Statistics* 23:ix-xxvii.

Kintanar, Roman L. 1984. *Climate of the Philippines.* Manila: Philippine Atmospheric, Geophysical, and Astronomical Services Administration.

Korten, David C. 1986. Comment. *Regional Development Dialogue* 7:156–60.

Krinks, Peter. 1983. Rectifying Inequality or Favouring the Few? Image and Reality in Philippine Development. In *Rural Development and the State,* edited by A. M. Lea and D. P. Chaudhri. London: Methuen.

_____. 1975. Changing Land Use on a Philippine Frontier. *Agricultural History* 49:473–90.

_____. 1974. Old Wine in a New Bottle: Land Settlement and Agrarian Problems in the Philippines. *Journal of Southeast Asian Studies* 5:1–17.

_____. 1970. Peasant Colonization in Mindanao. *Journal of Tropical Geography* 30:38–47.

Kumar, Raj. 1986. *The Forest Resources of Malaysia: Their Economics and Development.* Singapore: Oxford University Press.

Kwi, Soong Ngin; Haridas, G.; Seng, Yeoh Choon; and Hua, Tan Peng. 1980. *Soil Erosion and Conservation in Peninsular Malaysia.* Kuala Lumpur: Rubber Research Institute of Malaysia.

Lachowski, Henry M.; Dietrich, David; Umali, Ricardo; Aquino, Edgardo; and Basa, Virgilio. 1979. LANDSAT Assisted Forest Land-Cover Assessment of the Philippine Islands. *Photogrammetric Engineering and Remote Sensing* 45:1387–91.

Lampman, Robert J. 1967. Some Interactions between Economic Growth and Population Change in the Philippines. *Philippine Economic Journal* 6:1–20.

Lanly, J. P. 1985. Defining and Measuring Shifting Cultivation. *Unasylva* 37:17–21.

_____. 1982. *Tropical Forest Resources.* Rome: FAO.

Lansigan, Nicolas. 1959. An Appraisal of Forestry in the Philippines. *Filipino Forester* 2:4–50.

Leach, Gerald, and Mearns, Robin. 1988. *Beyond the Woodfuel Crisis.* London: Earthscan Publications.

Lewis, Laurence, and Coffey, William J. 1985. The Continuing Deforestation of Haiti. *Ambio* 14:158–60.

168   BIBLIOGRAPHY

Llapitan, Eduardo A. 1983. Agroforestry: The Bureau of Forest Development Experience. In Philippine Council for Agriculture and Resources Research and Development, *Agroforestry in Perspective*. Los Baños: PCARRD.

Lopez, Maria Elena. 1989. Personal communication.

Lopez-Gonzaga, Violeta. 1987. *Capital Expansion, Frontier Development, and the Rise of Monocrop Economy in Negros (1850-1898)*. Occasional Paper no. 1. Bacolod: LaSalle University.

Lugo, Ariel E., and Brown, Sandra. 1982. Conversion of Tropical Moist Forests: A Critique. *Interciencia* 7:89–93.

Lugo, Ariel E.; Schmidt, Ralph; and Brown, Sandra. 1981. Tropical Forests in the Caribbean. *Ambio* 10:318–24.

Luna, Cardozo M. 1982. Agricultural Productivity and Urbanization in the Philippines. M.A. thesis, University of the Philippines.

Luna, Telesforo W. 1966. The Land and Natural Resources of the Philippines. In *Proceedings of the First Conference on Population, 1965*, edited by Population Institute. Quezon City: University of the Philippines Press.

Luning, H. A. 1981. *The Need for Regionalized Agricultural Development Planning: Experiences from Western Visayas, Philippines*. College: Southeast Asian Regional Center for Graduate Study and Research in Agriculture.

Lynch, Owen J., and Talbott, Kirk. 1988. Legal Responses to the Philippine Deforestation Crisis. *Journal of International Law and Politics* 20:679–713.

Mackie, Cynthia. 1984. The Lessons behind East Kalimantan's Forest Fires. *Borneo Research Bulletin* 16:63–73.

Magno, Alexander R. 1985. The Marcos Regime: Crisis of Political Reproduction. In United Nations University Asian Perspectives Project (Southeast Asia), *Transnationalization, the State, and the People: The Philippine Experience*, part 2. Quezon City: University of the Philippines.

Mahar, Dennis A. 1989. *Government Policies and Deforestation in Brazil's Amazon Region*. Washington, D.C.: World Bank.

Makil, Perla Q. 1984. Forest Management and Use: Philippine Policies in the Seventies and Beyond. *Philippine Studies* 32:27–53.

Mangahas, Mahar. 1975. The Philippines. In Economic and Social Commission for Asia and the Pacific, *Comparative Study of Population Growth and Agricultural Change: A Survey of the Philippines and Thailand Data*. Asian Population Studies Series no. 23. Bangkok: United Nations,

Mangahas, Mahar, and Barros, Bruno. 1980. The Distribution of Income and Wealth: A Survey of Philippine Research. *Survey of Philippine Development Research* 1:51–132.

Markus, Gregory B. 1979. *Analyzing Panel Data*. Beverly Hills: Sage Publications.

Marx, U. 1985. *Area Development of Forest Types in the Quirino Province from 1969 to 1981*. Munich: Deutsche Forstinventur-Service.

McLennan, Marshall S. 1980. *The Central Luzon Plain: Land and Society on the Inland Frontier*. Quezon City: Alemars.

Meijer, Willem. 1973. Devastation and Regeneration of Lowland Dipterocarp Forests in Southeast Asia. *BioScience* 23:528–33.

Melillo, J. M.; Palm, C. A.; Houghton, R. A.; Woodwell, G. M.; and Myers, N. 1985. A Comparison of Two Recent Estimates of Disturbance in Tropical Forests. *Environmental Conservation* 12:37–40.

Metz, John J. 1989. An Outline of Wild Vegetation Use in Upland Nepal. Paper presented at the Annual Meeting of the Association of American Geographers, Baltimore, March 19–22.

Mijares, Tito A., and Nazaret, Francisco V. 1974. *The Growth of Urban Population in the Philippines and Its Perspective.* Manila: Bureau of Census and Statistics, Technical Paper no. 5.

Millikan, Brent H. 1988. The Dialectics of Devastation: Tropical Deforestation, Land Degradation, and Society in Rondonia, Brazil. M.A. thesis, University of California, Berkeley.

Mizukoshi, Mitsuhara. 1971. Regional Divisions of Monsoon Asia by Koppen's Classification of Climate. In *Water Balance of Monsoon Asia: A Climatological Approach,* edited by Mastatoshi Yoshino. Honolulu: University of Hawaii Press.

Molloy, Ivan. 1983. *The Conflicts in Mindanao: Whilst the Revolution Rolls on, the Jihad Falters.* Centre of Southeast Asian Studies Working Paper no. 30. Melbourne: Monash University.

Mydans, Seth. 1988. In Post-Marcos Philippines, Corruption Still a Way of Life. *New York Times* 808 (47661): 1, 10 (Oct. 17).

Myers, Norman. 1988a. Environmental Degradation and Some Economic Consequences in the Philippines. *Environmental Conservation* 15:205–14.

———. 1988b. Tropical Deforestation and Climate Change. *Environmental Conservation* 15: 293–98.

———. 1988c. Tropical Deforestation and Remote Sensing. *Forest Ecology and Management* 23: 215–25.

———. 1984. *The Primary Source.* New York: W. W. Norton.

———. 1980. *Conversion of Tropical Moist Forests.* Washington, D.C.: National Academy of Sciences.

Nair, C. T. S. 1985. Crisis in Forest Resource Management. In *India's Environment: Crisis and Responses,* edited by J. Bandyopadhyay, N. D. Jayal, V. Schoettli, and C. Singh. New Delhi: Natraj Publishers.

National Census and Statistics Office. 1985. *1980 Census of Agriculture: National Summary.* Manila: NCSO.

———. 1983. *Urban Population of the Philippines by Category, by Region, Province, and City/ Municipality and by Barangay: 1970, 1975, and 1980.* Special Report no. 4. Manila: NCSO.

———. 1981. Interregional Migration in the Philippines: 1970–1975. *Journal of Philippine Statistics* 32:vii–xv.

———. 1974. *1971 Census of Agriculture,* vol. 2, *National Summary.* Manila: NCSO.

———. n.d. *Philippines 1980; Population, Land Area, and Density: 1970, 1975, and 1980.* Manila: NCSO.

National Economic Council. 1959. *The Raw Materials Resources Survey: Series no. 1, General Tables.* Manila: Bureau of Printing.

National Economic Development Authority. 1987. *Philippine Statistical Yearbook, 1987.* Manila: NEDA.

———. 1982. *Regional Development: Issues and Strategies on Urbanization and Urban Development.* Regional Planning Studies Series no. 8. Manila: NEDA.

———. 1981. *Regional Development: Issues and Strategies on Agriculture.* Regional Planning Studies Series no. 3. Manila: NEDA.

———. 1980. *Philippine Statistical Yearbook, 1980.* Manila: NEDA.

National Environmental Protection Council. 1983. *The Philippine Environment, 1982.* Quezon City: Ministry of Human Settlements.

National Irrigation Authority. 1988. Service Area of National, Communal, and Pump Systems by Province, 1967–1987. Computer printouts.

National Remote Sensing Agency. 1983. *Mapping of Forest Cover in India from Satellite Imagery.* Hyderabad: Department of Space.

National Research Council. 1982. *Ecological Aspects of Development in the Humid Tropics.* Washington, D.C.: National Academy Press.

National Statistics Office. 1987. *Philippine Yearbook, 1987.* Manila: National Statistics Office.

National Water Resources Council. 1977. *Irrigation Inventory.* Manila: Republic of the Philippines.

Natural Resources Management Center. 1980. *Assessment of Forestry Management Problems and Issues: A Delphi Approach.* Quezon City: Ministry of Natural Resources.

——. 1979. *Philippine Natural Resources Statistics,* vol. 8, *Forestry.* Manila: Ministry of Natural Resources.

Nelson, Gerald C. 1984. *The Impact of Government Policies on Forest Resource Utilization.* Working Paper 84–04. Manila: Philippine Institute for Development Studies.

Nelson, Gerald C., and Cruz, Wilfrido. 1985. Macro Policies and Forestry. Paper prepared for the Agricultural Development Council Forestry Seminar, Sapporo, Japan, June.

Nguiagain, Titus. 1985. Trends and Patterns of Internal Migration in the Philippines, 1970–80. *Philippine Economic Journal* 24:234–62.

Nilsson, N. E.; Marsch, H. E.; and Singh, K. D. 1978. *Identification and Planning of a National Forest Inventory for the Philippines.* Rome: FAO/United Nations Development Programme.

Nordin, C. F., and Meade, R. H. 1982. Deforestation and Increased Flooding of the Upper Amazon. *Science* 215:426–27.

Ofreno, Rene E. 1980. *Capitalism in Philippine Agriculture.* Quezon City: Foundation for Nationalist Studies.

Olofson, Harold. 1984. Toward a Cultural Ecology of Ikalahan Dooryards: A Perspective for Development. *Philippine Quarterly of Culture and Society* 12:306–25.

——. 1983. Indigenous Agroforestry Systems. *Philippine Quarterly of Culture and Society* 11:149–74.

——. 1980. Swidden and *Kaingin* among the Southern Tagalog: A Problem in Philippine Upland Ethno-Agriculture. *Philippine Quarterly of Culture and Society* 8:168–80.

Osemeobo, Gbdebo Jonathan. 1988. The Human Causes of Forest Depletion in Nigeria. *Environmental Conservation* 15:17–28.

Oshima, Harry T. 1987. *Economic Growth in Monsoon Asia.* Tokyo: University of Tokyo Press.

Owen, Norman G. 1983. *Philippine-American Economic Interactions: A Matter of Magnitude.* In *The Philippine Economy and the United States: Studies in Past and Present Interactions,* edited by Norman G. Owen. Ann Arbor: University of Michigan, Center for South and Southeast Asian Studies.

Paderanga, Cayetano. 1986. *A Review of Land Settlement in the Philippines, 1900–1975.* School of Economics Discussion Paper no. 8613. Quezon City: University of the Philippines.

Palm, C. A.; Houghton, R. A.; and Melillo, J. M. 1986. Atmospheric Carbon Dioxide from Deforestation in Southeast Asia. *Biotropica* 18:177–88.

Palmier, Leslie. 1989. Corruption in the West Pacific. *Pacific Review* 2:11–23.

Palo, Matti. 1980. *Forest Sector Statistics: A Development Plan for the Philippines.* Manila: FAO.

Palo, Matti; Salmi, Jyrki; and Mery, Gerardo. 1987. Deforestation in the Tropics: Pilot Scenarios Based on Quantitative Analyses. In *Deforestation or Development in the Third World,* edited by Matti Palo and Jyrki Salmi. Helsinki: Finnish Forest Research Institute.

Panayotou, Theodore. 1983a. Present Status of Asian Tropical Forests and Needed Measures: An Overview. Paper presented at the joint ADC/Japan Seminar on the Management of Forest Resources: Issues of Forest Policy in the Developing Countries in Asia, Los Baños, July 10–14.

_____. 1983b. *Renewable Resource Management for Agriculture and Rural Development: Research and Policy Issues.* Bangkok: Agricultural Development Council.

Panayotou, Theodore, and Sungsuwan, Somthawin. 1989. *An Econometric Study of the Causes of Tropical Deforestation: The Case of Northeast Thailand.* Discussion Paper. Cambridge: Harvard Institute of International Development.

Pantastico, Eduvigis B., and Metra, Teodula M. 1982. Impact of Mining and Logging on Soil Resources in Some Places in the Philippines. In *Proceedings of the First International Symposium on Soil, Geology, and Landforms-Impact on Land Use Planning in Developing Countries,* edited by P. Nutalaya. Bangkok: Association of Geochemists for International Development.

Paper Industry Corporation of the Philippines. 1988. Personal communication.

Parsons, James J. 1976. Forest to Pasture: Development or Destruction? *Revisita de Biologica Tropical* 24:121–38.

Pelzer, Karl J. 1978. Swidden Cultivation in Southeast Asia: Historical, Ecological, and Economic Perspectives. In *Farmers in the Forest,* edited by Peter Kunstadter, E. C. Chapman, and Sanga Sabhasri. Honolulu: University of Hawaii Press.

_____. 1945. *Pioneer Settlement in the Asiatic Tropics.* Special Publication no. 29. New York: American Geographical Society.

_____. 1941. *An Economic Survey of the Pacific Area,* part 1, *Population and Land Utilization.* New York: Institute of Pacific Relations.

Perez, Aurora E. 1985. *Spatial Mobility and Development in Retrospect.* Quezon City: University of the Philippines, Population Institute.

Pernia, Ernesto M. 1988. Urbanization and Spatial Development in the Asian and Pacific Region: Trends and Issues. *Asian Development Review* 6:86–105.

Pernia, Ernesto M.; Paderanga, Cayetano W.; Hermoso, Victorina P.; and Associates. 1983. *The Spatial and Urban Dimensions of Development in the Philippines.* Manila: Philippine Institute for Development Studies.

Pernikar, Harald T. 1984. Adapting a Chainsaw Export-Import Company to the Changing Condition of the Philippine Forests. M.A. thesis, Ateneo de Manila University.

Philippine Council for Agriculture and Resources Research. 1981. *Data Series on Rice Statistics in the Philippines.* Los Baños: Philippine Council for Agriculture and Resources Research.

Philippine Council for Agriculture and Resources Research and Development. 1982. *The Philippines Recommends for Dipterocarp Production.* Los Baños: Philippine Council for Agriculture and Resources Research and Development.

Philippine Lumberman. 1985. Licenses of 11 Log Firms Cancelled: Bare "Salting" of $38 Million in 1984. 31 (4): 6–7.

_____. 1970. The Lumberman Outlook. 16 (5): 3.

_____. 1969. The Lumberman Outlook. 15 (10): 22.

Plumwood, Val, and Routley, Richard. 1982. World Rainforest Destruction: The Social Factors. *Ecologist* 12:4–22.

Poblacion, Gregorio. 1959. Logging in the Philippines. *Filipino Forester* 11:89–106.

Poore, Duncan. 1976. *Ecological Guidelines for Development in Tropical Rain Forests.* Morges: IUCN.

Population Institute, ed. 1966. *First Conference on Population, 1965.* Quezon City: University of the Philippines Press.

Population Reference Bureau, Inc. 1990. *World Population Data Sheet.* Washington, D.C.: Population Reference Bureau.

Population, Resources, Environment, and the Philippine Futures (PREPF). 1980. *Probing Our Futures: The Philippines 2000 A.D.* Manila: National Economic Development Authority.

Porter, Gareth D., and Ganapin, Delfin. 1988. *Resources, Population, and the Philippines' Future.* Washington, D.C.: World Resources Institute.

Power, John H., and Sicat, Gerardo P. 1971. *The Philippines: Industrialization and Trade Policy.* London: Oxford University Press.

Power, John H., and Tumaneng, Tessie D. 1983. *Comparative Advantage and Government Price Intervention Policies in Forestry.* Working Paper 83–05. Manila: Philippine Institute for Development Studies.

Presidential Economic Staff. 1968. *Country Report on the Philippine Forest-Based Industries.* Manila: Office of the President.

Primack, Richard B. 1988. Forestry in Fujian Province, People's Republic of China, during the Cultural Revolution. *Arnoldia* 48:26–29.

Proctor, John. 1985. Tropical Rain Forest: Ecology and Physiology. *Progress in Physical Geography* 9:402–13.

Pryor, Robin J. 1979. The Philippines: Patterns of Population Movement to 1970. In *Migration and Development in South-East Asia: A Demographic Perspective,* edited by Robin J. Pryor. Kuala Lumpur: Oxford University Press.

Ramos-Jimenez, Pilar; Chong-Javier, M. Elena; and Sevilla, Judy C. 1986. *Philippine Urban Situation Analysis.* Manila: UNICEF.

Reis, E. J., and Margulis, S. 1990. Economic Perspectives on Deforestation in the Brazilian Amazon. Paper presented at the Project Link Conference, Manila, Philippines, Nov. 5–9.

Repetto, Robert. 1990. Deforestation in the Tropics. *Scientific American* 262 (4): 36–42.

———. 1988. *The Forests for the Trees? Government Policies and the Misuse of Forest Resources.* Washington, D.C.: World Resources Institute.

Repetto, Robert, and Gillis, Malcolm, eds. 1988. *Public Policies and the Misuse of Forest Resources.* New York: Cambridge University Press.

Republic of the Philippines and United Nations Children's Fund. 1987. *Situation of Children and Women in the Philippines.* Manila: Republic of the Philippines/UNICEF.

Revilla, A. V. 1991. Forest Change Assessment in the NFI (Indonesia) Project. Paper presented at the FAO Workshop on Methodology for Deforestation Assessment in Southeast Asia, Bangkok, May 5–17.

———. 1988. The Constraints to and Prospects for Forest Development in the Philippines. In *Proceedings of the RP-German Forest Resources Inventory Application of Results to Forest Policy,* edited by Ralph Lennertz and Konrad Uebelhor. Quezon City: Forest Management Bureau.

———. 1987. Personal communication.

———. 1983. Policies and Strategies to Perpetuate the Philippine Dipterocarp Forests. Manuscript.

_____. 1981. Land Assessment and Management for Sustainable Uses in the Philippines. In *Assessing Tropical Forest Lands: Their Suitability for Sustainable Use*, edited by Richard Carpenter. Dublin: Tycooly.

Reynolds, Lloyd G. 1985. *Economic Growth in the Third World, 1850–1980*. New Haven: Yale University Press.

Richards, John F., and Tucker, Richard P., eds. 1988. *World Deforestation in the Twentieth Century*. Durham: Duke University Press.

Richards, John F.; Haynes, Edward S.; and Hagen, James R. 1985. Changes in the Land and Human Productivity in Northern India, 1870–1970. *Agricultural History* 59:523–48.

Robles, Alan C. 1987. Who Are Killing Our Forests? *Manila Chronicle (Focus, Sunday Magazine* July 5): 1–3.

Rocamora, J. Eliseo. 1979. Rural Development Strategies: The Philippine Case. In *Approaches to Rural Development: Some Asian Experiences*, edited by Inayatullah. Kuala Lumpur: Asian and Pacific Development Administrative Center.

Roche, Frederick C. 1988. Java's Critical Uplands: Is Sustainable Development Possible? *Food Research Institute Studies* 21:1–43.

Roque, Celso R. 1978. Remote Sensing Programs of the Philippines. *Proceedings of the Twelth International Symposium on Remote Sensing of Environment*, vol. 1, March 20–26. Ann Arbor: University of Michigan.

Rosenberg, Jean G., and Rosenberg, David A. 1980. *Landless Peasants and Rural Poverty in Indonesia and the Philippines*. Ithaca: Cornell University, Center for International Studies.

Roth, Dennis M. 1983. Philippine Forests and Forestry: 1565–1920. In *Global Deforestation and the Nineteenth Century World Economy*, edited by Richard P. Tucher and J. F. Richards. Durham: Duke University Press.

Rudel, Thomas K. 1989. Population, Development, and Tropical Deforestation. *Rural Sociology* 54 (3): 327–38.

Russell, Susan D. 1989. Simple Commodity Production and Class Formation in the Southern Cordillera of Luzon. Paper presented at the Annual Meeting of the Association of Asian Studies, Washington, D.C., March 17–19.

Sacerdoti, Guy, and Galang, Jose. 1985. The Seeds of Change. *Far Eastern Economic Review* 130:103–10 (Oct. 31).

Salvador, Eduardo; Argete, Eriberto; Versoza, Celso; Opena, Feliciano; del Rosino, Erlito; and Salac, Ronilo. 1985. *Update of Forest Inventory and Mapping of Ilocos Norte Using Aerial Photographs*. Research Monograph no. 2. Quezon City: Natural Resource Management Center.

Sanchez, M. Solita P. 1986. Illegal Trade in Logs: The RP-Japan Case. *CRC Staff Memos* 20:3–6.

Schade, Jurgen. 1988. Consequences of the FRI for Forest Policy. In *Proceedings of the RP-German Forest Resources Inventory Application of Results to Forest Policy*, edited by Ralph Lennertz and Konrad Uebelhor. Quezon City: Forest Management Bureau.

Scientific American. 1986. Seeing the Forest. 255:66–67.

Scott, Geoffrey A. J. 1979. The Evolution of the Socio-Economic Approach to Forest Occupance *(Kaingin)* Management in the Philippines. *Philippine Geographical Journal* 23:58–73.

Scott, Margaret. 1989. The Disappearing Forests. *Far Eastern Economic Review* 143 (2): 34–38 (Jan. 12).

Scotti, Roberto. 1990. *Estimating and Projecting Forest Area at Global and Local Level: A Step Forward*. Rome: FAO.

Segura, Marian; Revilla, A. V.; and Bonita, M. L. 1977. A Historical Perspective of Philippine Forest Resources. In PREPF (vol. 2). Manila: Development Academy of the Philippines.

Segura-de los Angeles, Marian. 1986. Economic Analysis of Resource Conservation by Upland Farmers in the Philippines. Ph.D. diss., University of the Philippines.

_____. 1985. Economic and Social Impact Analysis of an Upland Development Project in Nueva Ecija, Philippines. *Journal of Philippine Development* 12:324–94.

_____. 1982. Research on Forest Policies for Philippine Development Planning: A Survey. *Survey of Philippine Development Research* 2:3–55.

Segura-de los Angeles, Marian; Cruz, Cerenilla; and Corpuz, Eumelia. 1988. The Private Costs of Commercial Forestry, Reforestation, and Social Forestry. Paper presented at the Upland Resource Policy Conference, sponsored by the Philippine Institute for Development Studies and the Department of the Environment and Natural Resources, Quezon City, March 14.

Serevo, Tiburcio; Asiddao, Florencio; and Reyes, Martin. 1962. Forest Resources Inventory in the Philippines. *Philippine Journal of Forestry* 18:1–19.

Serna, Cirilio B. 1986. *Degradation of Forest Resources.* Bangkok: FAO.

Shahani, Leticia Ramos (Senator). 1988. A Moral Recovery Program: Building a People—Building a Nation. A report submitted to the Philippine Senate Committees on Education, Arts, and Culture and on Social Justice, Welfare, and Development, May 9.

Shirley, Hardy. 1960. Some Observations on Philippine Forestry with Special Emphasis on Forestry Education. Los Baños: University of the Philippines, College of Forestry.

Shrestha, Ramrajya L. 1986. *Socioeconomic Factors Leading to Deforestation in Nepal.* Research and Planning Paper Series no. 2. Morrilton, Ark.: Winrock International.

Simkins, Paul D., and Wernstedt, Frederick L. 1971. *Philippine Migration: The Settlement of the Digos-Padada Valley, Davao Province.* Southeast Asian Studies Monograph Series no. 16. New Haven: Yale University.

Singh, K. D. 1979. *Photo-Interpretation and Mapping System for the Philippine Forests.* Rome: FAO.

Sioli, Harald. 1985. The Effects of Deforestation in Amazonia. *Geographical Journal* 151:197–203.

Smith, Nigel J. 1976. Brazil's Transamazon Highway Settlement Scheme: Agrovilas, Agropoli, and Ruropoli. In *Proceedings of the Association of American Geographers,* vol. 8.

Smith, P. C. 1977. The Evolving Pattern of Interregional Migration in the Philippines. *Philippine Economic Journal* 16:121–59.

Society of Filipino Foresters. 1959–60. Comment on the "Gill Report." *Philippine Lumberman* 6.

Soerianegara, Ishemat. 1982. Socio-Economic Aspects of Forest Resources Management in Indonesia. In *Socio-Economic Effects and Constraints in Tropical Forest Management,* edited by E. G. Hallsworth. London: John Wiley and Sons.

Southeast Asia Chronicle. 1979. Tribal People and the Marcos Regime: Cultural Genocide in the Philippines. No. 67 (Oct.): entire issue.

Southgate, Douglas, and Pearch, David. 1988. *Agricultural Colonization and Environmental Degradation in Frontier Developing Economies.* Environment Department Working Paper no. 9. Washington, D.C.: World Bank.

Spencer, J. E. 1966. *Shifting Cultivation in Southeastern Asia.* Berkeley: University of California Press.

Sricharatchanya, Paisal. 1987. Jungle Warfare. *Far Eastern Economic Review* 137 (38): 86–88 (Sept.17).

Steinberg, David Joel. 1967. *Philippine Collaboration in World War II.* Manila: Solidaridad.

Stocking, Mike. 1987. Measuring Land Degradation. In *Land Degradation and Society*, edited by Piers Blaikie and Harold Brookfield. London: Methuen.

Stone, Richard L. 1971. "Lagay" and the Policeman: A Study of Private Transitory Ownership of Public Property. In *Modernization: Its Impact in the Philippines*, vol. 5, edited by Frank Lynch and Alfonso de Guzman. IPC Papers no. 10. Manila: Ateneo de Manila University Press.

Sulit, Carlos. 1963. Brief History of Forestry and Lumbering in the Philippines. *Journal of the American Chamber of Commerce of the Philippines* 39:16–24.

_____. 1947. Forestry in the Philippines during the Japanese Occupation. *Philippine Journal of Forestry* 5:22–47.

Sungsuwan, Somthawin. 1985. A Study on the Causes of Deforestation in Northeast Thailand. M.A. thesis, Thammasat University.

Susman, Paul; O'Keefe, Phil; and Wisner, Ben. 1983. *Global Disasters: A Radical Interpretation*. In *Interpretations of Calamity*, edited by K. Hewitt. Boston: Allen and Unwin.

Sutlive, Vinson H.; Altshuler, Nathan; and Zamora, Mario D., eds. 1981a. *Where Have All the Flowers Gone? Deforestation in the Third World*. Studies in Third World Societies Publication no. 13. Williamsburg: College of William and Mary.

_____. 1981b. *Blowing in the Wind: Deforestation and Long-Range Implications*. Studies in Third World Societies Publication no. 14. Williamsburg: College of William and Mary.

Swan, Bernard. 1979. Geology, Landforms, and Soils. In *South-East Asia: A Systematic Geography*, edited by R. D. Hill. Kuala Lumpur: Oxford University Press.

Swedish Space Corporation. 1988. *Mapping of the Natural Conditions of the Philippines*. Solna: Swedish Space Corporation.

Talbot, Lee M., and Talbot, Martha A. 1964. *Renewable Natural Resources in the Philippines: Status, Problems, and Recommendations*. Manila: IUCN.

Tamesis, Florencio. 1963. Philippine Forests and the Lumber Industry. In *Shadows on the Land*, edited by Robert E. Huke. Manila: Bookmark.

_____. 1956. The Philippines. In *A World Geography of Forest Resources*, edited by Stephen Haden-Guest, John K. Wright, and Eileen M. Teclaff. New York: Ronald Press Co.

_____. 1948. Philippine Forests and Forestry. *Unasylva* 6:316–25.

_____. 1937. *General Information on Philippine Forests*. Manila: Bureau of Printing.

Tapales, Proserpina Domingo. 1986. Assessing the Impact of Authoritarian Rule on the Philippine Government: What Efficiency? What Accountability? *Pilipinas* 6:21–34.

Tesoro, Flor. 1989. *Wood Supply and Demand Policy Study*. Manila: Department of Environment and Natural Resources/Development Academy of the Philippines.

Tharan, Sri. 1976. Systems Corruption in Revenue Raising: The Loss of Timber as a Source of State Revenue in Kelantan (1960–1974). In *IRDC (Asia) Project on Bureaucratic Behavior and National Development*. Ottawa: International Development Research Center.

Thung, Heng L. 1972. An Evaluation of the Impact of a Highway on a Rural Environment in Thailand by Aerial Photographic Methods. Ph.D. diss., Cornell University.

Tiglao, Rigaberto. 1981. *The Philippine Coconut Industry*. Davao: ARC Publications.

Tiongzon, Mari Luz; Regalado, Aurora; and Pascual, Ramon. 1986. *Philippine Agriculture in the 70s and 80s: TNC's Boon, Peasants' Doom*. Agricultural Policy Studies no. 2. Quezon City: Philippine Peasant Institute.

Totman, Conrad. 1989. *The Green Archipelago: Forestry in Preindustrial Japan*. Los Angeles: University of California Press.

Tucker, Richard P., and Richards, J. F., eds. 1983. *Global Deforestation and the Nineteenth Century World Economy*. Durham: Duke University Press.

Uebelhor, Konrad. 1988. Impacts of FRI Results on Dipterocarp Forest Management. In *Proceedings of the RP-German Forest Resources Inventory Applications of Results to Forest Policy*, edited by Ralph Lennertz and Konrad Uebelhor. Quezon City: Forest Management Bureau.

Uhlig, Harald, ed. 1984. *Spontaneous and Planned Settlement in Southeast Asia*. Hamburg: Institute of Asian Affairs.

Ulack, Richard. 1977. Migration to Mindanao: Population Growth in the Final Stage of a Pioneer Frontier. *Tijdschrift voor Economische en Sociale Geografie* 68:133–44.

Ullman, Edward L. 1960. Trade Centers and Tributary Areas of the Philippines. *Geographical Review* 50:203–18.

Umali, Ricardo M. 1981. Forest Land Assessment and Management for Sustainable Uses in the Philippines. In *Assessing Tropical Forest Lands: Their Suitability for Sustainable Uses*, edited by Richard A. Carpenter. Dublin: Tycooly.

U.S. Agency for International Development. 1980. *Preliminary Analysis of Philippine Poverty as a Base for a U.S. Assistance Strategy*. Manila: USAID.

U.S. Bureau of the Census. 1905. *Census of the Philippine Islands, 1903*. Washington, D.C.: Government Printing Office.

U.S. Department of State. 1980. *Draft Environmental Report on the Philippines*. Washington, D.C.: Department of State.

Utleg, Juan L. 1970. Logging and the Problem of Forest Destruction. *Philippine Lumberman* 8:30–32.

_____. 1958. Land Classification in the Philippines. *Philippine Journal of Forestry* 14:33–58.

Vandermeer, Canute. 1963. Corn Cultivation on Cebu: An Example of an Advanced Stage of Migratory Farming. *Journal of Tropical Geography* 17:172–77.

Vandermeer, Canute, and Agaloos, Bernardo C. 1962. Twentieth Century Settlement of Mindanao. *Papers of the Michigan Academy of Science, Arts, and Letters* 47.

Van Oosterhout, A. 1983. Spatial Conflicts in Rural Mindanao, the Philippines. *Pacific Viewpoint* 24:29–49.

Walker, Robert T. 1987. Land Use Transition and Deforestation in Developing Countries. *Geographical Analysis* 19:18–30.

_____. 1985. *Behavioral Models of Deforestation*. Regional Research Institute Paper no. 8605. Morgantown: West Virginia University.

Watts, Michael. 1983. On the Poverty of Theory: Natural Hazards Research in Context. In *Interpretations of Calamity*, edited by K. Hewitt. Boston: Allen and Unwin.

Weidelt, Hans J., and Banaag, Valeriano S. 1982. *Aspects of Management and Silviculture of Philippine Dipterocarp Forests*. Eschborn: German Agency for Technical Cooperation.

Wernstedt, Frederick L., and Spencer, J. E. 1967. *The Philippine Island World*. Berkeley: University of California Press.

Wernstedt, Frederick L., and Simkins, Paul D. 1965. Migration and the Settlement of Mindanao. *Journal of Asian Studies* 25:83–103.

Westoby, Jack. 1989. *Introduction to World Forestry*. Oxford: Basil Blackwell.

_____. 1981. Who's Deforesting Whom? *IUCN Bulletin* 14:124–25.

Whitford, H. N. 1911. *The Forests of the Philippines*. 2 vols. Bulletin no. 10. Manila: Bureau of Forestry.

Whitlow, J. R. 1980. *Deforestation in Zimbabwe: Problems and Prospects*. Salisbury: University of Zimbabwe.

Whitmore, T. C. 1984. Vegetation Map of Malesia at Scale of 1:5,000,000. *Journal of Biogeography* 11:461–71.

Willett, Kip. 1976. *Resource Planning in the Philippines: An Interim Report.* Washington, D.C.: International Bank for Reconstruction and Development.

Williams, Robert G. 1986. *Export Agriculture and the Crisis in Central America.* Chapel Hill: University of North Carolina Press.

Wong, J., and Arief, S. 1984. An Overview of Income Distribution. *Southeast Asian Economic Review* 5:1–44.

Woodwell, G. M.; Hobbie, J. E.; Houghton, R. A.; Melillo, J. M.; Peterson, B. J.; Shaver, G. R.; Stone, T. A.; Moore, B.; and Park, A. B. 1983. *Deforestation Measured by LANDSAT: Steps toward a Method.* Washington, D.C.: U.S. Department of Energy.

World Bank. 1989a. *Philippines: Environment and Natural Resource Management Study.* Washington, D.C.: World Bank.

_____. 1989b. *Philippines: Toward Sustaining the Economic Recovery.* Washington, D.C.: World Bank.

_____. 1978. *Forestry Sector Paper.* Washington, D.C.: World Bank.

_____. 1976. *The Philippines: Priorities and Prospects for Development.* Washington, D.C.: World Bank.

World Resources Institute. 1990. *World Resources, 1990–91.* Washington, D.C.: World Resources Institute.

World Resources Institute, World Bank, and United Nations Development Programme. 1985. *Tropical Forests: A Call for Action.* Washington, D.C.: World Resources Institute.

Wurfel, David. 1983. The Development of Post-War Philippine Land Reform: Political and Sociological Explanations. In *Second View from the Paddy*, edited by Antonio Ledesma, Perla Makil, and Virginia Miralao. Manila: Ateneo de Manila University Press.

_____. 1979. The Changing Relationship between Political and Economic Elites in the Philippines. In *Poverty and Social Change in Southeast Asia*, edited by Ozay Mehmet. Ottawa: University of Ottawa Press.

Wyatt-Smith, John. 1987. Problems and Prospects for Natural Management of Tropical Moist Forests. In *Natural Management of Tropical Moist Forests*, edited by François Mergen and Jeffrey R. Vincent. New Haven: Yale University, School of Forestry and Environmental Studies.

Zon, Raphael. 1910. *The Forest Resources of the World.* Washington, D.C.: Government Printing Office.

Zon, Raphael, and Sparhawk, William N. 1923. *Forest Resources of the World.* New York: McGraw Hill.

Zosa-Feranil, Imelda. 1988. Personal communication.

_____. 1987. *Persisting and Changing Patterns of Population Redistribution in the Philippines.* Quezon City: University of the Philippines, Population Institute.

Index

THE UNIVERSITY OF CHICAGO
GEOGRAPHY RESEARCH PAPERS
(Lithographed, 6 x 9 inches)

Titles in Print

127. GOHEEN, PETER G. *Victorian Toronto, 1850 to 1900: Pattern and Process of Growth.* 1970. xiii + 278 p.

131. NEILS, ELAINE M. *Reservation to City: Indian Migration and Federal Relocation.* 1971. x + 198 p.

132. MOLINE, NORMAN T. *Mobility and the Small Town, 1900-1930.* 1971. ix + 169 p.

133. SCHWIND, PAUL J. *Migration and Regional Development in the United States, 1950-1960.* 1971. x + 170 p.

134. PYLE, GERALD F. *Heart Disease, Cancer and Stroke in Chicago: A Geographical Analysis with Facilities, Plans for 1980.* 1971. ix + 292 p.

136. BUTZER, KARL W. *Recent History of an Ethiopian Delta: The Omo River and the Level of Lake Rudolf.* 1971. xvi + 184 p.

139. McMANIS, DOUGLAS R. *European Impressions of the New England Coast, 1497-1620.* 1972. viii + 147 p.

142. PLATT, RUTHERFORD H. *The Open Space Decision Process: Spatial Allocation of Costs and Benefits.* 1972. xi + 189 p.

143. GOLANT, STEPHEN M. *The Residential Location and Spatial Behavior of the Elderly: A Canadian Example.* 1972. xv + 226 p.

144. PANNELL, CLIFTON W. *T'ai-Chung, T'ai-wan: Structure and Function.* 1973. xii + 200 p.

145. LANKFORD, PHILIP M. *Regional Incomes in the United States, 1929-1967: Level, Distribution, Stability, and Growth.* 1972. x + 137 p.

148. JOHNSON, DOUGLAS L. *Jabal al-Akhdar, Cyrenaica: An Historical Geography of Settlement and Livelihood.* 1973. xii + 240 p.

149. YEUNG, YUE-MAN. *National Development Policy and Urban Transformation in Singapore: A Study of Public Housing and the Marketing System.* 1973. x + 204 p.

150. HALL, FRED L. *Location Criteria for High Schools: Student Transportation and Racial Integration.* 1973. xii + 156 p.

151. ROSENBERG, TERRY J. *Residence, Employment, and Mobility of Puerto Ricans in New York City.* 1974. xi + 230 p.

152. MIKESELL, MARVIN W., ed. *Geographers Abroad: Essays on the Problems and Prospects of Research in Foreign Areas.* 1973. ix + 296 p.

154. WACHT, WALTER F. *The Domestic Air Transportation Network of the United States.* 1974. ix + 98 p.

160. MEYER, JUDITH W. *Diffusion of an American Montessori Education.* 1975. xi + 97 p.

162. LAMB, RICHARD F. *Metropolitan Impacts on Rural America.* 1975. xii + 196 p.

163. FEDOR, THOMAS STANLEY. *Patterns of Urban Growth in the Russian Empire during the Nineteenth Century.* 1975. xxv + 245 p.

164. HARRIS, CHAUNCY D. *Guide to Geographical Bibliographies and Reference Works in Russian or on the Soviet Union.* 1975. xviii + 478 p.

165. JONES, DONALD W. *Migration and Urban Unemployment in Dualistic Economic Development.* 1975. x + 174 p.

166. BEDNARZ, ROBERT S. *The Effect of Air Pollution on Property Value in Chicago.* 1975. viii + 111 p.

167. HANNEMANN, MANFRED. *The Diffusion of the Reformation in Southwestern Germany, 1518-1534.* 1975. ix + 235 p.

168. SUBLETT, MICHAEL D. *Farmers on the Road: Interfarm Migration and the Farming of Noncontiguous Lands in Three Midwestern Townships. 1939-1969.* 1975. xiii + 214 p.

169. STETZER, DONALD FOSTER. *Special Districts in Cook County: Toward a Geography of Local Government.* 1975. xi + 177 p.

172. COHEN, YEHOSHUA S., and BRIAN J. L. BERRY. *Spatial Components of Manufacturing Change.* 1975. vi + 262 p.

173. HAYES, CHARLES R. *The Dispersed City: The Case of Piedmont, North Carolina.* 1976. ix + 157 p.

174. CARGO, DOUGLAS B. *Solid Wastes: Factors Influencing Generation Rates.* 1977. 100 p.

176. MORGAN, DAVID J. *Patterns of Population Distribution: A Residential Preference Model and Its Dynamic.* 1978. xiii + 200 p.

177. STOKES, HOUSTON H.; DONALD W. JONES; and HUGH M. NEUBURGER. *Unemployment and Adjustment in the Labor Market: A Comparison between the Regional and National Responses.* 1975. ix + 125 p.

180. CARR, CLAUDIA J. *Pastoralism in Crisis. The Dasanetch and Their Ethiopian Lands.* 1977. xx + 319 p.

181. GOODWIN, GARY C. *Cherokees in Transition: A Study of Changing Culture and Environment Prior to 1775.* 1977. ix + 207 p.

183. HAIGH, MARTIN J. *The Evolution of Slopes on Artificial Landforms, Blaenavon, U.K.* 1978. xiv + 293 p.

184. FINK, L. DEE. *Listening to the Learner: An Exploratory Study of Personal Meaning in College Geography Courses.* 1977. ix + 186 p.

185. HELGREN, DAVID M. *Rivers of Diamonds: An Alluvial History of the Lower Vaal Basin, South Africa.* 1979. xix + 389 p.

186. BUTZER, KARL W., ed. *Dimensions of Human Geography: Essays on Some Familiar and Neglected Themes.* 1978. vii + 190 p.

187. MITSUHASHI, SETSUKO. *Japanese Commodity Flows.* 1978. x + 172 p.

188. CARIS, SUSAN L. *Community Attitudes toward Pollution.* 1978. xii + 211 p.

189. REES, PHILIP M. *Residential Patterns in American Cities: 1960.* 1979. xvi + 405 p.

190. KANNE, EDWARD A. *Fresh Food for Nicosia.* 1979. x + 106 p.

192. KIRCHNER, JOHN A. *Sugar and Seasonal Labor Migration: The Case of Tucumán, Argentina.* 1980. xii + 174 p.

194. HARRIS, CHAUNCY D. *Annotated World List of Selected Current Geographical Serials, Fourth Edition. 1980.* 1980. iv + 165 p.

196. LEUNG, CHI-KEUNG, and NORTON S. GINSBURG, eds. *China: Urbanizations and National Development.* 1980. ix + 283 p.

197. DAICHES, SOL. *People in Distress: A Geographical Perspective on Psychological Well-being.* 1981. xiv + 199 p.

198. JOHNSON, JOSEPH T. *Location and Trade Theory: Industrial Location, Comparative Advantage, and the Geographic Pattern of Production in the United States.* 1981. xi + 107 p.

199-200. STEVENSON, ARTHUR J. *The New York–Newark Air Freight System.* 1982. xvi + 440 p.

201. LICATE, JACK A. *Creation of a Mexican Landscape: Territorial Organization and Settlement in the Eastern Puebla Basin, 1520-1605.* 1981. x + 143 p.

202. RUDZITIS, GUNDARS. *Residential Location Determinants of the Older Population.* 1982. x + 117 p.

204. DAHMANN, DONALD C. *Locals and Cosmopolitans: Patterns of Spatial Mobility during the Transition from Youth to Early Adulthood.* 1982. xiii + 146 p.

206. HARRIS, CHAUNCY D. *Bibliography of Geography. Part II: Regional. Volume 1. The United States of America.* 1984. viii + 178 p.

207-208. WHEATLEY, PAUL. *Nagara and Commandery: Origins of the Southeast Asian Urban Traditions.* 1983. xv + 472 p.

209. SAARINEN, THOMAS F.; DAVID SEAMON; and JAMES L. SELL, eds. *Environmental Perception and Behavior: An Inventory and Prospect.* 1984. x + 263 p.

210. WESCOAT, JAMES L., JR. *Integrated Water Development: Water Use and Conservation Practice in Western Colorado.* 1984. xi + 239 p.

211. DEMKO, GEORGE J., and ROLAND J. FUCHS, eds. *Geographical Studies on the Soviet Union: Essays in Honor of Chauncy D. Harris.* 1984. vii + 294 p.

212. HOLMES, ROLAND C. *Irrigation in Southern Peru: The Chili Basin.* 1986. ix + 199 p.

213. EDMONDS, RICHARD LOUIS. *Northern Frontiers of Qing China and Tokugawa Japan: A Comparative Study of Frontier Policy.* 1985. xi + 209 p.

214. FREEMAN, DONALD B., and GLEN B. NORCLIFFE. *Rural Enterprise in Kenya: Development and Spatial Organization of the Nonfarm Sector.* 1985. xiv + 180 p.

215. COHEN, YEHOSHUA S., and AMNON SHINAR. *Neighborhoods and Friendship Networks: A Study of Three Residential Neighborhoods in Jerusalem.* 1985. ix + 137 p.

216. OBERMEYER, NANCY J. *Bureaucrats, Clients, and Geography: The Bailly Nuclear Power Plant Battle in Northern Indiana.* 1989. x + 135 p.

217-218. CONZEN, MICHAEL P., ed. *World Patterns of Modern Urban Change: Essays in Honor of Chauncy D. Harris.* 1986. x + 479 p.

219. KOMOGUCHI, YOSHIMI. *Agricultural Systems in the Tamil Nadu: A Case Study of Peruvalanallur Village.* 1986. xvi + 175 p.

220. GINSBURG, NORTON; JAMES OSBORN; and GRANT BLANK. *Geographic Perspectives on the Wealth of Nations.* 1986. ix + 133 p.

221. BAYLSON, JOSHUA C. *Territorial Allocation by Imperial Rivalry: The Human Legacy in the Near East.* 1987. xi + 138 p.

222. DORN, MARILYN APRIL. *The Administrative Partitioning of Costa Rica: Politics and Planners in the 1970s.* 1989. xi + 126 p.

223. ASTROTH, JOSEPH H., JR. *Understanding Peasant Agriculture: An Integrated Land-Use Model for the Punjab.* 1990. xiii + 173 p.

224. PLATT, RUTHERFORD H.; SHEILA G. PELCZARSKI; and BARBARA K. BURBANK, eds. *Cities on the Beach: Management Issues of Developed Coastal Barriers.* 1987. vii + 324 p.

225. LATZ, GIL. *Agricultural Development in Japan: The Land Improvement District in Concept and Practice.* 1989. viii + 135 p.

226. GRITZNER, JEFFREY A. *The West African Sahel: Human Agency and Environmental Change.* 1988. xii + 170 p.

227. MURPHY, ALEXANDER B. *The Regional Dynamics of Language Differentiation in Belgium: A Study in Cultural-Political Geography.* 1988. xiii + 249 p.

228-229. BISHOP, BARRY C. *Karnali under Stress: Livelihood Strategies and Seasonal Rhythms in a Changing Nepal Himalaya.* 1990. xviii + 460 p.

230. MUELLER-WILLE, CHRISTOPHER. *Natural Landscape Amenities and Suburban Growth: Metropolitan Chicago, 1970-1980.* 1990. xi + 153 p.

231. WILKINSON, M. JUSTIN. *Paleoenvironments in the Namib Desert: The Lower Tumas Basin in the Late Cenozoic.* 1990. xv + 196 p.

232. DUBOIS, RANDOM. *Soil Erosion in a Coastal River Basin: A Case Study from the Philippines.* 1990. xii + 138 p.

233. PALM, RISA, AND MICHAEL E. HODGSON. *After a California Earthquake: Attitude and Behavior Change.* 1992. xii + 130 p.

234. KUMMER, DAVID M. *Deforestation in the Postwar Philippines.* 1992. xviii + 179 p.